design by people **for people**
ESSAYS ON
USABILITY

Edited by:
Russell J. Branaghan

design by people **for people**
ESSAYS ON
USABILITY

Edited by:
Russell J. Branaghan

USABILITY PROFESSIONALS' ASSOCIATION

Table of Contents

Conceptual Modeling

Web Site Evaluation

Preface | Russell J. Branaghan, Fitch Inc.

The past ten years have introduced revolutionary changes in how we stay informed, conduct business, stay healthy, keep in touch, and entertain ourselves. Typically, these changes dash from the research and development laboratories to prominent roles in offices, homes, classrooms, transportation vehicles, and hospitals.

Unfortunately, technologies change much quicker than people do. It takes a long time to change behavior, beliefs, habits, work processes, and cultures. And there are clear limitations on what we can expect people to do—how quickly they can learn, what they can remember, what text size they can read, and so on. Ignoring these issues has drastic consequences for commerce, government, and safety.

As I write this introduction, we are still unsure who the next president of the United States will be. The election took place almost a month ago, but the ballots in Florida were confusing to many voters. As a result, some voters who intended to vote for Al Gore actually cast a vote for Pat Buchanan. Some of these voters then recognized their mistake and tried to correct it, resulting in a double-punched ballot which cannot be counted. It is not clear how many of these votes would have gone to Gore, but poor usability may very well have cost him the presidency.

For the past decade the Usability Professionals' Association (UPA) has focused on making these technologies and their resulting products and services safer, easier to learn, more efficient to use, and generally more satisfying. The group is unique in that it focuses on the everyday application of usability principles, methods, and techniques rather than on esoteric theory. The members of this society shared their techniques and experiences in their newsletter *Common Ground*.

Over the years, *Common Ground* developed into a useful and substantial source for usability information. This book assembles some of its best contributions in one document so that they may serve as a reference and introduction to usability. It covers topics ranging from when to conduct each usability activity during a product's development to how to be a better consultant. I believe it will be helpful to anyone who influences the usability of products, services, or documents.

The UPA Publications Board members, especially Cindy Clark, Elizabeth Rosenzweig, and Charlie Kreitzberg, have been of great assistance in shaping this document. JoAnn Hackos and her colleagues were very helpful in providing us with the original texts. Several employees at Fitch made large contributions also: Cindy Chischilly designed the cover art and the page layout, Jean Konrad assisted in editing the book, and Sheri Worrall was in charge of production. Ashley Menges, Brian Aubert, Jason Plumb, and Marylyn Wright worked on collecting and formatting the material.

Finally, I would like to thank the contributors to this book—the practitioners who work to make life better and safer for people in these changing times. Their willingness to share their work, knowledge, and experience with others made this a document that will influence design for a long time to come.

Russell J. Branaghan, Ph.D.

03 December 2000

Columbus, Ohio

A Letter from the Founder

Janice James
Founder and Past President, UPA
Simply Usable Through Design

My prediction still stands: "The Usability Professionals' Association will remain for a long while a vital association for usability professionals." Those were my words in 1996 when, in *Common Ground,* I thought about the early years of UPA.

As I think back to where we were as a profession when we held our first meeting as a SIG at CHI '91 with about 50 attendees, I have to conclude that we've come a long way. Through our individual efforts as usability professionals and through the publication of our insights and experiences, we've made a difference in corporations, the products they develop, and in the general public's understanding and awareness about usability. Our association, UPA, has successfully held eight conferences, published a quarterly newsletter since 1991, and has a membership of over 1,600 usability professionals. Just in the past year, UPA has established three chapters, including one in Europe.

I have to think we've definitely made progress.

On the other hand, when I finish phone conversations like the one I just had with another usability professional who is struggling because his company "just doesn't get it," I sometimes feel that we still have so far to go as a profession and in what we need to accomplish as a professional association. The good news is that those stories seem to occur less often today than in years past. More frequently, I hear reports of the start-up of a new human factors/usability group in a corporation—and sometimes even the restart of a previously abolished group!

One of the major goals of UPA is to provide a network and

opportunities through which usability professionals can communicate and share information about skills and skill development, methodologies, tools, technology, and organizational issues. From the beginning, the focus has been on practical vs. strictly theoretical information. We've continued to work toward this goal through the Association's newsletter, *Common Ground,* and through our annual conferences, as well as the networking that takes place between members everyday through mediums such as Utest.

I've witnessed a number of changes in the practice of usability engineering methods over the last twelve years. But there still seems to be a real mix in how "usability" is positioned within corporations and a continued large effort to figure out where the perfect fit is and how to get management to recognize its value. I guess that will never change, although I'd like to think that it will steadily improve.

Unfortunately, within some corporations, I've seen a real separation/distinction of efforts between usability testing, user interface design, and ethnographic studies. More frequently, however, I've seen a blending of these professions and professionals to create more hybrid usability practices. Professionals in the fields of technical communication, anthropology, psychology, human factors, visual/graphic and interaction design, and marketing are practicing commonly recognized usability engineering methods and adding a slant of emphasis/practice from their own original backgrounds and methodologies.

Overall, there also seems to be much less formal and time-intensive usability methods being practiced. I think this is largely due to the overall greater sense of urgency and competitiveness that the Internet has brought to our lives in recent years.

Luckily, no matter how much state-of-the-art technological developments affect our usability practices, a constant set of core methods, experiences, and information transcends time and remains valuable to us from year to year. This book is a good example of that. In looking back at all the past issues of *Common Ground,* I realized how valuable the issues remain. Don Ballman's article in the February/March 1995 issue of *Common Ground,* titled "How Well Do We Know Our Users? Alternate Ways of Viewing Test Participants" is a good example. After having just completed a usability study for a client, his article made me think about some of the user types (introverts vs. extroverts) who participated in my recent study. Don made me think more deeply about the ways in which the participants behaved during the study, the level of feedback they shared, and how I might rethink my recruiting process for user participants in the future to achieve richer and possibly more valid feedback.

This book is, in a sense, a tribute to UPA and especially to *Common Ground* and how it has exemplified the Association's importance and value over the past eight years. In addition to the many contributing authors, many of whom are represented in this book, is the group of editors who through hard work and dedication made *Common Ground* possible. Peter Mitchell, in 1991, named and provided us our first important issues of *Common Ground*. A year later, Don Ballman took over Peter's efforts, and successfully expanded the publication's offerings for the next fours years until JoAnn Hackos assumed the editor position in 1996. JoAnn introduced a wonderful, new, and updated look to the newsletter and provided UPA an excellent, professional newsletter through 1999.

In my final reflections about the Association and its many facets, I'm taken back to the beginning years. Some key people come to mind who were instrumental in the early shaping and growth of UPA. The more I think, the more names I come up with who have been so dedicated to UPA year after year. But, when I focus on the first very early years of UPA, I can't help but remember Kay Chalupnik and Dave Rinehart. They both spent endless hours with me after our first informal meeting at CHI '91 until two years later when the three of us, as founding board members, finally signed the bylaws and incorporated the Association in 1993. Without Kay and Dave's efforts, as well as those of Jack Young and Margaret Beier, UPA wouldn't have survived its beginning.

As I look past those early years to the future, I think about the many growth possibilities for UPA. One of those that is essential is of course for UPA to become a more global organization. From the beginning, UPA has always been successful in attracting participation from usability professionals from many countries outside of the United States. With an international board member representative joining the board in 1998 and the establishment of European and Canadian chapters in 1999, UPA is on its way to becoming more internationally focused.

This book recognizes many more people who, through contributions to issues of *Common Ground,* have assisted UPA in its growth. It's a book to keep through your professional career. It won't become outdated—it's a classic so to speak. The information contained in these chapters will be helpful to you now, just as it was when it was first written, and just as it will be in years to come.

Janice James

Founder and Past President, UPA

Making Computer and Internet Usability a Priority

Charles B. Kreitzberg
and Ben Shneiderman,
Cognetics Corporation

As usability professionals, we are all too aware of the productivity losses, frustration, and lost business that result from poorly designed user interfaces. And we are uncomfortable with the risks created by poorly designed computer systems in life-critical applications such as air travel, medical care, and military applications.

Yet despite the common sense of our approach, we still find it difficult to convince the technical and managerial communities that usability is a critical business parameter. While senior managers may support the concept of usability, project managers and developers, coping with too tight schedules, often see it as a nicety that can be eliminated.

As business transitions into the new economy, usability has become a strategic business goal. In "new corporations," enterprisewide systems reach out to consumers for sales and service and to vendors and strategic partners, as well. Corporations that cannot deliver easy-to-use systems will find it hard to compete in an increasingly competitive marketplace. And competition is not only coming from long-established companies, but from new, upstart companies that are unburdened by the need to retread or replace complex legacy systems.

Users, too, face a great deal of frustration. They still feel guilty when they make mistakes and often feel that they should somehow be able to figure out what to do. The fact that IDG's "for dummies" series generated $121 million in revenue last year suggests something about how users view themselves as well as their hunger to master the technology. Unfortunately, poor design, both of the software UI and user assistance, makes it difficult for them and wastes tens of billions of dollars in lost productivity.

Over time, it is likely that market forces will encourage more attention to software usability. In e-commerce, for example, consumers will evaluate the quality of their interactive shopping experience much as they evaluate their in-store experiences. Companies that produce confusing or rigid interfaces will lose customers. A hint of that market pressure was reported by the Boston Consulting Group in a study released in April 2000 that found that a full 28 percent of online purchasing transactions fail. And consumers are angry. Twenty-eight percent of consumers who suffered a failed purchase attempt stopped shopping online; 23 percent stopped purchasing at the site in question; and 6 percent also stopped patronizing the retailer's physical store.

But while market forces will ultimately force improvements in the user experience, corporations would do far better to take an active role in promoting usability than waiting for evolution to run its uncertain course. As business transitions to the information economy, those who insult, frustrate, and poorly serve their customers' needs will pay a heavy price. We need to establish usability as a priority now. And having gotten the attention of management, we need to teach the software industry the techniques of user-centered design. Let's do it right!

Crafting a message

As an industry, usability professionals need to present a clear, consistent message. We propose that it should contain the following three elements:

1. Good usability is good business

2. Poor usability is a failure of management

3. Correcting the problem is straightforward and well within the scope of normal business practice.

The way in which we deliver the message must be pointed and unambiguous. While among ourselves we will surely continue to debate nuances, outside the usability community we must deliver punch and clarity. The message needs to be crafted so it speaks directly to managers, software developers, and end users. It needs to be supported by data that support the business case. And as an industry, we need a better public relations pipeline to get the message to the media.

With a clear message in hand, we need to look at the reasons that usability has not been a higher priority in the business community. In our opinion, there are three pillars of development that will support the needed change: (1) increased technology fluency in the business community, (2) a cultural shift in IT, and (3) the integration of user-centered design into product development life cycles.

Technology fluency in the business community

Computer technology is complex and often confusing to the uninitiated. Even information technology professionals often find themselves struggling to understand technology that seems inevitably rapidly changing, fragmented, immature, and poorly documented. Imagine the pain of the non-technical end user.

While making users feel empowered is a significant concern, we are equally concerned by the negative impact that uninformed users can have on the quality of design. Line managers who are not fluent in technology are often intimidated. Increasingly they are taking on fiscal responsibility for projects that they barely understand. And because they don't know how to make good decisions, they often defer to technologists without fully understanding what they are agreeing to. This fact was painfully obvious at a recent meeting in which an IT representative met with a senior executive and vendor to help evaluate the vendor's proposal. Midway in the meeting, the IT "professional" stalked out, telling the others that they were wasting his time and didn't understand anything. The senior executive confided that she was so dependent upon the IT department that she could not fight back. "We're terrified of them," she confided.

Perhaps in the past, "leaving it to IT" was acceptable, but business today cannot afford the luxury of losing the insight and wisdom of its business professionals. Software that will be used by consumers, sales, and service staff must reflect a profound understanding of the products and services being offered and how to position them effectively. This is not a job for which most programmers are qualified.

When the partnership between the business and IT is not effective, the usability of the software inevitably suffers. Participatory design cannot be effective if business partners cannot express their needs or vision alternative solutions. In the anecdote we related above, the obvious problem was the rude and unprofessional behavior of the IT representative. While such behavior is clearly unacceptable, we see a deeper issue in the senior executive's lack of technology fluency. Without the conceptual knowledge to understand the proposal, she was unable to employ her excellent business skills to manage the decision-making process.

Most users have by now mastered the basic operation of their computers. They can operate a word processor, use e-mail, create a spreadsheet, and get on to the Web. But technology fluency is more than this. A 1999 report by the National Research Council, "Being Fluent with Information Technology," argues that familiarity with a few basic software programs "is too modest a goal in the presence of rapid change…." The report suggests that a higher level of competence— technology fluency—is required so that individuals can vision and apply technology to their work and personal lives. We do not think that business managers and business professionals should become programmers. We *do* think that they should become fluent in the language and concepts of technology.

Usability professionals, as with business analysts, have often seen an aspect of our jobs as compensating for our business clients' lack of technological sophistication. But the role of "interpreter" is no longer enough in the fast-moving transition to e-business. We would like to see usability professionals help business managers and professionals develop more technology fluency. In addition to helping with design and evaluation, usability professionals can also serve as technology coaches, helping business professionals acquire the concepts and skills necessary to become full partners in the development process.

We understand that some usability professionals may disagree with the idea that business users should have to understand technology. They would like to protect users from the details. We see many examples in other fields, however, where an educated understanding of the basics is a powerful tool. It is hard to make a responsible decision about repairing your car if you cannot have a meaningful conversation with the mechanic. And getting good medical care is almost impossible for patients who do not understand how to evaluate their options. Our vision of the user-centered future is one in which senior executives, line managers, and business professionals are able to "think in the language of technology" and function as full partners to their information technology counterparts.

Creating a cultural shift in IT

Cultural artifacts have a long life. In the early days of computing, the information technology department was a self-contained unit which maintained the hardware, developed software, and ran the programs. Business units requested "program runs" and received the output, but had little direct responsibility for technology.

In this environment, it was assumed that engineers and programmers would manage all aspects of software; the user interfaces were designed for technical users. The pressure for efficient use of machine resources led to decisions that put heavy cognitive burdens on the users, but they were often drawn in to the challenge of dealing with complexity. In the 1970s, for example, running a program on an IBM mainframe required the user to construct a set of job control language (JCL) commands like the following:

```
//QUICK JOB, 'JOE USER,' TIME=(0,5)
//JOBLIB DD DSN=MWD.DT34A.LOADLIB,
//   DISP=SHR
//STEP1 EXEC PGM=IEBGENER,REGION=1024K
//UPFILE DD DSN=QWL.DS34B.BKUP.MSTRBKUP(+00),
//   DISP(OLD,KEEP,KEEP)
//INFILE DD DSN=QWL.DS34B.BKUP.MSTRBKUP(-01),
//   DISP(OLD,KEEP,KEEP),
//   UNIT=(AFF=MSTRBKUP)
//SYSIN DD *
//
```

JCL was meaningful only to an information technology professional and it was a rare user who would brave it. For an IT professional, however, investing the time to learn JCL was reasonable. In short, the technology world before the pc was inaccessible to most users.

The personal computer, of course, changed the "social" environment, but a lot of cultural baggage from the past still remains. When the personal computer emerged, the need for easier-to-learn, easier-to-use, and easier-to-remember software interfaces became apparent. The desktop GUI metaphor of the Xerox STAR in 1981 and the Apple Macintosh in 1984 suggested a direction that would enable less-technical people to use computers. There were fewer esoteric commands to learn, typing errors were dramatically reduced, and even intermittent users could remember what to do from session to session.

When Microsoft adopted the desktop GUI metaphor for Windows, it seemed as if significant progress in user interface design might be made. In the past 15 years, as successive versions of Windows came to dominate the market, there has been inadequate progress in usability. This has led to the paradox of more people doing more work on computers, without basic improvement in UI design. It remained for the emergence of the World Wide Web to once again change the social equation.

The Web has transformed the computer into a mass medium like the television or telephone. While business users participating in design require a high level of technical fluency, consumers cannot be relied upon to have fluency. The simple point-and-click hypertext interface of the World Wide Web is a good starting point, but far too simplistic to support full-featured Web applications. As Web programming tools such as Java mature, we can expect the emergence of complex Web applications. As Web applications grow in complexity, the associated user interfaces will likely become more complex, as well. Without a commitment to user-centered design, we can expect Web usability to become an even more serious problem than the client-server software of the past decade.

If business is to be successful in satisfying this mass market of consumers, there needs to be a realization on the part of software developers that to support millions of users, the industry must cast off the cultural remnants of the old mainframes. IT professionals must come to respect usability as a valued asset of their products. And they must come to understand that they, as information professionals, have a far different mental model than their target users. One of the most valuable aspects of usability testing is that it is objective and empirical. It does not rely on hand-waving arguments about what users can do or what users want. It easily answers the question of what works and what doesn't.

Our vision of the user-centered future is one in which information technology professionals have more appreciation of the "soft" skills that business professionals bring and understanding the value of incorporating business wisdom into the products they create. We hope that information technology professionals will understand that they operate within a complex, technical world and that they must restrain the technology from bleeding into the interfaces of their products. And, perhaps most of all, they must understand that interaction design cannot fall victim to the stress of production. It is not a nicety, but a necessity.

User-centered design

Usability professionals clearly understand the value of user-centered design. Few outside the profession do. This leads to conflict and misunderstanding. Particularly frustrating is the lack of attention to UI design in object-oriented analysis and design (OO/AD) that frequently takes the position that UI design should be deferred to the end of the software development cycle rather at the beginning. This is, in our opinion, a flawed assumption and as OO/AD becomes standard in many development shops, user-centered design is placed at risk.

There is a consensus among usability professionals as to the steps that compose an acceptable framework for user-centered design. This consensus is reflected in various commercial methodologies such as the LUCID Framework and the ISO standard 13407 Human-Centered Design Process For Interactive Systems.

Design and validation of the user interface is a small part of the overall software (or product) development process. But it is an essential one. Our vision of the future is that user-centered design process will be incorporated into all software development methodologies, ensuring that user-related issues are considered at appropriate points throughout the development process and guaranteeing a usable result.

The Fate of Software Projects

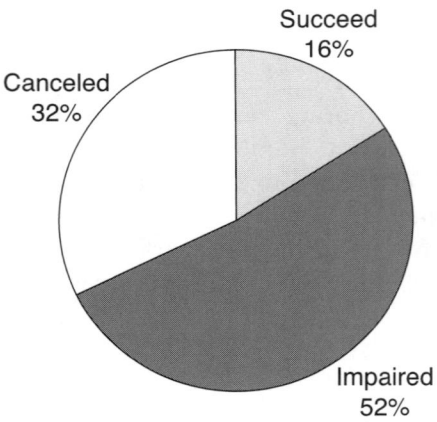

Measuring the losses

Although every usability professional has seen the impact of poorly designed user interfaces, we have too few studies that quantify the problem. Because sales of personal computers have been so important a part of the growing information economy, few analysts have challenged the assumption that more computers equaled more productivity. Stephen Roach of Morgan Stanley, a skeptic of the value of information technology, has been examining the "productivity paradox" which found little broad support for the thesis that information technology increases productivity. In a May 1998 interview with *CIO* magazine, Roach said, "We spend a lot and get little from it . . . the problem is, a lot of indiscriminate spending on relatively low-value-added functionality in the IT and the universe. I think the paybacks remain decidedly disappointing."

Roach has not been alone in lamenting information technology productivity. Paul Strassman, in his 1997 book, *The Squandered Computer,* said, "Computers are only tools. They are not an unqualified blessing. . . . They enhance sound business practices. They also aggravate inefficiencies whenever the people who use them are disorganized and unresponsive to customers' needs. The best computer technologies will always add unnecessary costs to a poorly managed firm. The problem seems to rest not with the inherent capabilities of the technologies, which are awesome, but with the managerial inability to use them effectively." Strassman cites a Gartner Group study that claims, "Seventy percent of IT projects have not delivered their expected benefits because they have failed to integrate the results into work processes."

The Standish Group reported similar numbers in a 1995 study titled "Chaos." The study surveyed 365 IT executive managers. The major finding was that only 9 to 16 percent of software development projects can be classified as successful, where success is defined as being completed on time, on budget, and implementing the original vision. Of the remainder, 31.1 percent of projects are canceled before they ever get completed, and 52.7 percent of projects will cost 189 percent of their original estimates. In terms of original vision, the study found that projects completed by the largest American companies have only approximately 42 percent of the originally proposed features. For smaller companies the number is a more reassuring 75 percent.

The study noted that the numbers related only to the cost of development but that the lost opportunity cost could be much higher. Of particular interest to usability professionals, is the Standish Group's analysis of the factors that predict success in software development projects. The top ten factors are shown in the table below along with weighting indicating their relative importance. Most of these factors are components addressed in user-centered design approaches.

SUCCESS CRITERIA	POINTS
1. User involvement	19
2. Executive management support	16
3. Clear statement of requirements	15
4. Proper planning	11
5. Realistic expectations	10
6. Smaller project milestones	9
7. Competent staff	8
8. Ownership	6
9. Clear vision & objectives	3
10. Hard-working, focused staff	3
TOTAL	**100**

From the usability professional's point of view, a key question is separating the inefficiencies due to poor UI design from other sources of inefficiency in information technology. Again, we have less research than would be desirable.

SBT Accounting Systems in San Rafael, California, conducted a 6,000-person survey of office workers and found non-productive use averages of 5.1 hours per week. SBT estimates the cost to American businesses is some $100 billion a year in lost productivity.

Tom Landauer's 1995 economic analysis, "The Trouble with Computers," suggested that applying user-centered design strategies to software would yield productivity gains of 40 to 80 percent. Translating this into economic terms could have a significant gain for the entire economy on the order of 3 percent.

Creating awareness

It is our belief that it is time for the usability profession to work together to make software usability (and other applications of usability, as well) a national priority. This means that business leaders, technical leaders, and end users will all recognize the value of well-designed software interfaces and will work together to solve the management problems that lead to poor designs. The field of usability or human-computer interaction (HCI), needs to be seen as a valid and essential subdiscipline of the computer industry, and the professional approach to usability engineering will lay the foundations for a change in priorities and practices in software development.

As a field, we can trace our origin to the historic conference in Gaithersburg, Maryland, in 1982 that many feel launched HCI as a distinct discipline. Since 1982, there has been a steady growth of research that ties computer science with psychology, graphic design, technical writing, and other topics. Today, we have at least four professional organizations with overlapping interests in promoting this field. Since 1991 the Usability Professionals' Association has promoted user-centered design approaches in business and government, supporting usability testing, expert reviews, and participatory design. ACM's Special Interest Group on Computer Human Interaction (SIGCHI) brings researchers and professionals together in an annual conference. The Human Factors & Ergonomics Society (HFES) and the Society for Technical Communication (STC) have also been active in promoting high-quality UI design.

Until now, however, these groups have addressed a fairly narrow audience and have been seen as peripheral (at best) within the software engineering community. Usability-oriented groups have taken a back seat to such topics as object-oriented programming, component-based software, and client-server networks.

By and large, the profession has been unsuccessful in creating awareness and action about the need for usability. So long as business managers feel unable to confront technology professionals, so long as users feel that they are responsible for their confusion, so long as IT executives do not acknowledge their responsibility to create highly usable tools, the situation will not change.

Media attention about usability issues is sporadic. Books such as Landauer's *The Trouble with Computers*, and Dertouzos's *What Will Be: How the World of Information Will Change Our Lives* get modest attention and journalists raise the issue from time to time, yet the issue does not receive sustained attention. But the media finds other issues far more exciting than end user struggles with software. Even when aircraft disasters (such as Korean Air 007 and the American Airlines crash in Cali or the AEGIS design flaw that contributed to the shoot-down of Iran Air 655) are determined to be related to user interface problems, there is little awareness that things could be different.

We will have sustained progress only when business fully understands that it is in their interest to promote the usability of their products: when there is public outcry, pressure from industrial buyers, and the government mandates requirements for publicly funded systems. Public awareness to promote change is primary. And it is up to the profession to make the case.

A call to action

It is time for the professional societies to extend their reach and initiate a sustained and vigorous campaign to promote awareness of the value of usability and the role of the usability professional. Through research and professional development, we have laid the foundation for action and for change, but our influence remains limited to a narrow community. Now we need to send our message to the larger community in terms that will promote action.

This is the right time to extend our efforts. Computing is no longer a private corporate activity. The Web has become central to delivery of software and has fundamentally changed the nature of computing. Through the Web, corporations reach out to their customers through e-commerce, and to their suppliers through enterprise resource planning (ERP) systems. Corporations have far less control over users in other organizations than they do over their own employees. Poorly designed user interfaces will make it difficult for the new corporation to achieve its goals and compete successfully.

As corporations become increasingly global, the role of the virtual workgroup becomes more important. Corporations need to locate talent which can be applied to a specific project, and initiate and manage the projects through computer-based communications tools. Here, too, high-quality UIs are critical to productivity.

Finally, the role of the IT department is evolving. There is increased recognition that user needs are not being met although changing in some corporations. Increasingly, aspects of software development responsibility are moving from a centralized IT department to the business units while IT retains responsibility for technical infrastructure.

How should such a program be structured? Given that the four professional societies have strong overlapping interests in promoting usability, a cooperative outreach program would be potentially effective. Such a program would enable them to pool scarce resources and also increase the impact of the message.

The outreach program needs to be multifaceted. Some possible components would include the following:

1. Conduct studies and report results to the media

The best way to get media attention is to conduct studies that quantify the problems caused by poor UI design. As a profession we have both the resources (usability laboratories) and expertise to document the problem. By doing so, we will be able to secure media attention and convince the business community and the government of the importance of our mission.

In addition to studies sponsored by the professional associations, we should also consider seeking funding from foundations and the government to document the extent of the problem and its economic consequences.

2. Develop technical training programs for usability professionals

While many usability professionals are technically adept, others have come to the profession with less technical backgrounds. To serve the role of user advocate best, usability professions must be able to meet software developers on their own ground. This does not mean that usability professionals need to become programmers, but it does suggest a need for a comprehensive conceptual foundation in software engineering and an understanding of emerging trends and problems. We suggest that the professional societies consider sponsoring technical training as part of a continuing education program for usability professionals.

3. Develop usability training programs for technical professionals

The complementary side of the training for usability professionals is providing software professionals with training in user-centered design. Until programmers understand our field, they will not respect it nor will they participate in it. Many of the books on UCD that have been written to date are focused on the usability professional. We suggest that the professional societies help develop courses and curricula for software engineers.

4. Develop technology fluency training for the business community

The third leg of the training effort is to help end users develop conceptual-level skills in technology and user-centered design so that they can become full participants in the design process. Despite the numerous opportunities for computer training that are available in the marketplace, this type of training is not widely available.

5. Work with software engineering methodology projects to ensure that UCD processes are incorporated

There is a lot of current work related to the development of new software engineering methodologies. In general, user-centered design is not emphasized in these methodologies and usability professionals rarely have a role in designing them. If user-centered design is not incorporated into these development methodologies, there is significant risk that it will be treated superficially in projects that follow the methodologies. If we are able to make UCD a part of software development methodologies, it will be easier to argue for its incorporation into development projects and the allocation of significant resources to it.

6. Work with the government to define procurement regulations

Finally, we believe that the usability professional should work with government to create mandates for minimum usability standards when software is procured for government use. This would apply both to commercial, off-the-shelf software and custom development projects.

7. Develop awards programs

Awards can be helpful in identifying good practices and products. Awards, such as those given by the American Institute of Architects, could influence other designers and industry leaders. Awards are useful stimulants of discussion and hopefully offer incentives to designers of the next generation of user interfaces.

Evaluations by such groups as Consumer's Union and Underwriters Laboratory can carry considerable weight. These groups maintain objectivity and independence that generate a level of trust in their evaluations. It would be useful for the professional organizations to explore working with such groups to establish minimum criteria for usability and promote public awareness.

Another area for potential alliances is for industry associations, such as the Software Publishers Association, that might find it in their own interest to become champions of usability.

Looking to the future

This is a critical time for the usability industry. The world of computing is changing and if we take a strong and coherent stand for user-centered design, we can emerge as key players in the development process. If we fail to do this, the role of the UI designer and evaluator may fall to less-skilled and -passionate players. As usability professionals, we need to raise our profile to the public, to software developers and to managers. We need to communicate the value of our position and teach the techniques we have developed. Only then will we be able to create tools that can transform society.

Communicating the Role of Usability Engineering | Paul McInerney, IBM

Usability practitioners expend much energy arguing with programmer/analysts that usability engineering is a distinct yet integral part of systems development in addition to the parts covered in traditional systems-development methodologies. This experience can be frustrating for both parties because of the gulf between disciplines. Hix and Hartson (1993, p. 10) characterize this gulf as follows: "People in these two worlds have different goals, attitudes, skills, perspectives, philosophies, needs, techniques, and tools. . . ." Despite the difficulties, explaining the role of usability remains an important goal: without its recognition by programmer/analysts, usability practitioners will be marginalized in systems development.

I recently faced this gulf when I delivered a guest lecture on the role of usability engineering to a university class on systems analysis and design. This article describes how I analyzed the communication barriers and addressed them in the lecture. This experience can point to some lessons for other practitioners communicating the role of usability to their project teams.

Overview of a systems-centered methodology

The course presented a traditional systems-centered methodology for developing business computer applications using the text by Whitten, et al, (1994). The methodology divides systems development into five phases—planning, analysis, design, implementation, and support. It also describes an information system as being composed of the following components or building blocks:

- People: the users, managers, and developers of information systems

- Data: the data and its organization in a database

- Activities: business activities and software applications that process data

- Networks: the distribution of people, data, activities, and technology to useful locations

- Technology: the hardware and software that supports the other building blocks

For each of the five phases of development, the methodology covers project activities related to each building block, with an emphasis on data and activities.

Communications barriers

In reviewing the text to prepare for the lecture, I identified several barriers that were going to make it tough to sell the role of usability.

- Competing claims about usability: Usability engineering and traditional systems development each claim they address usability.

- Ambiguous terminology: Usability engineering and traditional systems development both use some of the same terms but ascribe different meanings to them.

- Competing paradigms: Methodologies such as the one in the textbook portray themselves as addressing all system development issues; there is nothing missing including usability engineering.

- Limited audience interest: Most programmer/analysts (or usability professionals) have a limited interest in studying the relationship between usability engineering and the rest of systems development.

These barriers are discussed below, using the course text for illustration.

Competing claims about usability

The argument for usability engineering starts with the premise that traditional systems development does not adequately consider usability as a distinct concern. Usability professionals believe that their skills and techniques make a contribution to system usability beyond that of programmer/analysts. However, students who had read their textbook would have every reason to protest; they could point to several counterpoints such as —

• The text presents a list of ten principles for successful systems, with number one being: Know thy user.

• An entire chapter is devoted to user interface ("UI") design; it includes guidelines and right-sounding statements such as "too many systems are difficult to use because they exhibit poor human engineering" (p. 489).

Don't these points demonstrate that this methodology incorporates sufficient concern for usability? As this example shows, usability engineering and systems development have competing claims about who achieves usability. In fact, it is difficult to find anyone promoting a product, service, or methodology who does not claim a usability benefit. So, the first hurdle to overcome is the reaction, "Usability?! Of course our methodology (or product or service) addresses that. Why do we need usability engineering?"

Ambiguous terminology

Any attempt to introduce usability engineering quickly gets bogged down by terminology. Traditional systems development and usability engineering use the same words to mean different things, sometimes differing only in connotation or emphasis. Consider, for example, the terms "user" and "user requirements." To usability professionals, "user" refers mostly to the person using the system, while to programmers/analysts, "user" is a broader concept—anyone not on the development team, including the client or a representative of the worker (e.g., their boss) or a job title (e.g., office manager), and only rarely an actual person who will use the system directly.

Similarly, the term "user requirements" has connotations of task performance, cognitive information processing, and so on, to the usability community. To the programmer/analyst, it is more likely to mean business requirements (e.g., number of office locations where the system will be used) or business rules (e.g., verify that the date entered is a weekday). This terminology ambiguity further fogs any cross-disciplinary communication.

Competing paradigms

Another hurdle to arguing the role of usability engineering is that the students have learned and adopted the perspective of the systems-development methodology presented in their text. Many are also experienced programmers/analysts taking the course for continuing education. They are not passively receiving this message without preconceptions.

The text, as with any methodology, intends to describe all the pieces needed to develop a successful system and show how they fit together. The methodology is a complete framework—nothing missing and nothing superfluous. Those who have bought into a methodology tend to see anything new from that perspective, so it becomes difficult to convince them that something is missing or is wrong with it. Such a shortfall would call into question the whole edifice.

The text provides a good example of how a methodology can make it difficult to view systems development with any other perspective. Its five components (people, activities, data, networks, and technology) don't include user interface or usability engineering as a distinct component, although aspects are scattered within the other components. The authors reinforce their five-component perspective by displaying it as a pyramid—four sides are each labeled with a component and the fifth, technology, serving as the bottom of the pyramid. With this visualization, it is not possible to change the number of components without destroying the whole—which makes the pyramid a compelling representation of an information system.

Limited audience interest

Usability engineering can also be a hard sell because many people have limited interest in anything that sounds too abstract or methodological. This course was a required core course, which further suggested a limited audience interest in my guest lecture topic. As difficult as it was to engage this audience, usability practitioners know it can be even more challenging in a working environment where receptivity to change can be low in the face of project pressures. I heard of a meeting where a project manager told the usability engineering specialist, "I've heard enough about your process and theories—let's just get this project done!"

This lack of appreciation for the importance of process and methodology in developing a successful system is not limited to usability engineering. A programmer/analyst once told me that at his company the first thing to get thrown out the window when the product shipping deadline approaches is their systems-development process. The project team is told to just "do what it takes to meet the deadline." This is not an environment receptive to a message about the importance of usability engineering.

Approach used in the guest lecture

With these barriers in mind, I sought a way to convince the class that their textbook methodology should be modified to incorporate usability engineering. My approach included two techniques:

1. Build on what the students knew and accepted (i.e., their textbook) rather than argue they should discard it in favor of another approach

2. Present the argument graphically for maximum impact.

The results are shown in the tables below, taken from the lecture overheads. Table 1 summarizes (very briefly) the textbook methodology (described earlier); Table 2 presents a modification to incorporate usability engineering. Table 1 lays the groundwork for Table 2 in a couple of ways. First, it depicts the methodology as a table rather than as a pyramid; while presenting the same information, the table removes the implication that systems development includes exactly five components (one for each face of the pyramid and another for the base). In the table, each information-system building block is presented in a column and each phase of development is shown as a row.

Second, Table 1 depicts where the user interface design activity occurs in the textbook methodology. The UI is shown as being derived from the design of the functional components (i.e., data and activities); the UI design does not fit neatly into one column, visually suggesting that something is wrong with this approach.

Table 2 presents an improved methodology with user interaction/usability engineering as a distinct component, analogous to the other ones leading the design of the user interface. The improvement is evident from the appearance of the table; the UI now fits into one column, like the other components. Once the audience accepts the introduction of this new information-system component, they are ready to accept that this component must be addressed at each phase of development; they already accept this for the other components.

Conclusion

This article has identified some barriers to communicating the role of usability and has shown an attempt to overcome them. How was the message received? I was encouraged by some thoughtful comments and questions. However, I was reminded that the students are interested in the bottom line when someone asked, "Is this material on the final exam?"

TABLE 1. Traditional Systems Development
The UI is derived from the Process and Data Components

Information System Component

Phase of Development		Process	Data	Other Components
Planning		...	...	...
Analysis		business activity analysis essential DFD*	business data analysis essential ERD**	...
Design		implementation DFD program definition & design user interface	normalized ERD database definition & design	...
Implementation		(application programs & UI software)	databases	...
Support		...	...	...

*   DFD – Data Flow Diagram
** ERD – Entity Relation Diagram

TABLE 2. Systems Development with Usability Engineering
The UI is addressed as a distinct component at each phase of development

Information System Component

Phase of Development	User Interaction/ Usability Eng.	Process	Data	Other Components
Planning	...	...	...	...
Analysis	task analysis user profile, etc.	business activity analysis essential DFD*	business data analysis essential ERD**	...
Design	user interface design ↓↑ usability testing	implementation DFD program definition & design	normalized ERD database definition & design	...
Implementation	user interface software	application programs	databases	...
Support		...	...	...

* DFD – Data Flow Diagram
** ERD – Entity Relation Diagram

References

Hix, D. and H.R. Hartson. *Developing User Interfaces: Ensuring Usability Through Product and Process.* New York: John Wiley, 1993.

Whitten, J.L., Bentley and V. Barlowl. *Systems Analysis and Design Methods,* 3rd Ed. Boston: Irwin, 1994.

Experience Design

Interview with A. Henderson — Usability: from Operation to Empowerment

JoAnn Hackos,
ComTech Services

What issues should we be addressing when we consider the usability of technology? We have tended to focus on the technology. Instead, we must find ways to enable and empower the user by studying the complete context in which the user works.

Austin Henderson, manager of the Discourse Architecture Laboratory (part of Apple Research Laboratories), opened his keynote address at the 1996 UPA Conference at Copper Mountain, Colorado, by demonstrating through discussion and example just how important it is for us to move beyond the immediate operational needs of our users. The user community, he argues, is much broader than the operator of the hardware or software alone. The community includes the individual, the collaborators, the neighbors, and the society. All must be accounted for to achieve a usable design.

Operation: putting technology to use
Enablement: supporting operation
Empowerment: realizing enablement

Figure 1. Operation, Enablement, Empowerment.

Beginning in the early '80s, Austin and his team at Xerox PARC used ethnographic methods to study the community of users affected by the Xerox 8200. They learned that the interface and documentation that the trained operators loved were very difficult for casual users to use. Their investigation became iterative, with the perspective of the researchers constantly changing as they gained new insights (figure 2).

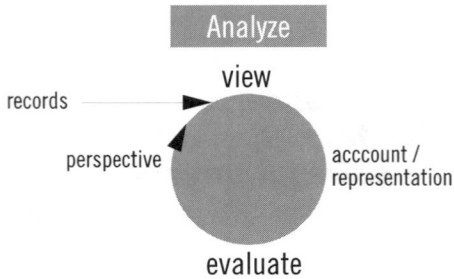

Figure 2. Iterative observation and analysis process.

Austin described the first product of iterative observation and analysis as a recognition that people have trouble performing tasks. They have trouble getting started, proceeding with their work, changing directions, and knowing when to stop. Technology, they learned, had to support the users' trouble, enabling them to get out of ordinary and inevitable problems easily. Austin concluded that usable technology must be able to:

• know what it's doing

• give the user an account of what's happening

• be prepared to change directions

• engage the user in a discourse about the work.

In the early '90s, Austin and his team began research on a system they called direct office sharing, a prototype of which they ran in their offices at Xerox PARC. They moved from analysis to a better understanding of what they called iterative design research. Figure 3 illustrates the continual relationship between design and analysis.

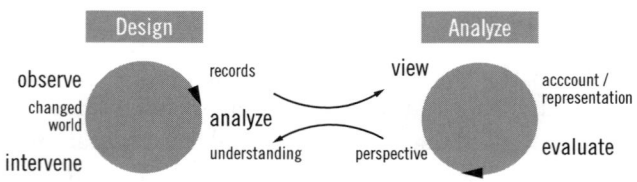

Figure 3. Iterative Design Research (after Scott Minneman).

The product of their efforts was a recognition that they had to move beyond the operator to an expansive view of usable technology. They postulated that effective technology was to be used by

- the operator (individual)

- collaborators (interactional)

- neighbors (communal)

- the institution (societal).

Simple user actions, such as making copies with the copy machine, involved a whole community of activity, from knowing how to perform a task and teaching others, to managing resources and developing new ways of working. Austin's continuing observation and analysis, now conducted at Apple, demonstrate that understanding the community of people who interact with a technology is much more than simply studying operation. We need to understand how people do what they do, to evaluate the impact of technology on the practices of the community, to know what rules the community establishes for itself to manage its interactions with technology. Technology is about people— not about "users."

Austin does not want us to view technology as value-neutral. Technology provides power by enabling users to define who makes the technology happen. It provides feelings of ownership and responsibility within a sociotechnical community. He would hope for a view that not only enables the operator to perform a task, but also empowers that operator and others in the community to negotiate with the technology and become, in a sense, a developer of the technology and its interactions. The supplier (product developer) develops a fixed technology but empowers the user to develop the rest of the interactions.

Austin ended his presentation with his pitch. He told the audience that professionals concerned with ensuring that technology is usable must address not merely the development of technology but the empowerment of people. They need to broadly enable people to employ technology in the continuing and continuous creation of socio-technical practice to meet whatever purposes they choose.

As designers, we need to listen to the user reflect upon the activity pursued. Then, we need to design technology that can itself engage in the user's dialogue. We must go beyond the problems we can anticipate to finding a way for the technology and the user to manage unanticipated problems. The current generation of software does not allow this dialogue to occur. An entirely new type of interface must be created, something that those in advanced research may now be working on. We have no way at present for our technology to infer what the user is trying to do, even through tracking sequences of key strokes. If we can engage the user in a discourse with the technology, perhaps we will find a means of enabling solutions for unanticipated problems. Until then, we continue watching rather than talking, pursuing the methods of ethnographic research to understand the user's world.

From Ease of Use to Experience Design

Russ Branaghan,
Fitch Inc.

I have a favorite old pair of sneakers that I can't seem to part with. They were white when I bought them, and occasionally, after a good washing, they return to white. However, now they are old and torn, and each washing reveals a new hole. My wife begs me to use them exclusively for mowing the lawn. Wearing them anywhere else is strongly discouraged.

Why in the world would I be so loyal to a pair of sneakers? Simply this: I bought them at the Nike Town store during the 1996 Summer Olympics in Atlanta. The store was a makeshift structure built exclusively for the Olympics, and it occupied at least one sizeable city block. Inside, shoppers could shoot baskets on the basketball court, attempt to hit tennis balls served with the velocity (or ferocity) of Pete Sampras, examine Michael Johnson's gold spikes, get autographs from current and past Olympic stars, or participate in many other activities related to sports. It was inspirational.

Of course, I bought much more than I planned, and the memory of that day prevents me from throwing any of those items out, least of all my Nike Airs. Nike skillfully designed an *event*, not a store, that I would remember for years to come. They didn't do it by designing a thing or a product—they did it by creating an experience. Everything in that store was designed with the sole (pun intended) purpose of inspiring couch potatoes, like me, to become athletes. Nike implemented a key principle that designers often forget—don't design things, design *experiences*. As Tom Peters urges, design the whole enchilada.

Business is beginning to value this perspective. For example, when choosing a restaurant for dinner, you can select from the jungle adventure provided by the Rainforest Café, the celebrity experience at Planet Hollywood, or the rock-and-roll experience at the Hard Rock Café. The primary offer of these restaurants is not food but an interesting adventure; the food is really secondary.

Retailing has often supplemented its product offers with experiences. For example, bookstores have long sponsored readings and autographs by their authors. But, now the experiences are more elaborate. My favorite sporting goods store has a climbing wall for adventurous types to practice their rock climbing. Many people go to the store with the intention of simply climbing the wall but walk out with a bag of sporting goods. Marketing professionals call this practice "entertailing." It's retailing and it's entertainment—it's an experience. And this experience pays off. On average, customers spend more than 45 minutes in Nike Town stores and as much as two hours in REI outfitters.

Hotel chains such as Marriott and Starwood are starting to design boutique hotels which offer smaller but more unique and interesting rooms for the customer. Not surprisingly, rooms like this fetch a handsome rate and spur a great deal of conversation. The key to these rooms is that they are interesting and memorable.

Movies, games, and amusement parks have always been on the leading edge of creating experiences, but recently even these industries have risen to a new level. A good example is the Imax Theater which boasts a screen that is nine stories high and 90 feet wide. I saw the movie *T-Rex* there, which required very sophisticated 3-D glasses to insert the viewer into the story. They worked. I felt like a part of the activity on the screen. During another Imax visit, I saw a documentary called *Speed* which was so realistic, I had severe motion sickness from watching the movie. Now that was an experience!

Recently, several companies have been working on interactive movie technology in which the characters respond differently depending on the verbal responses of the audience. Different audiences would actually experience different things in the same movie.

Virtual reality attempts to design real experiences, using visual and kinesthetic senses to immerse the user in the adventure. In terms of actual reality, adventure vacations represent one of the fastest-growing types of vacations these days. These vacations offer experiences in which the vacationer can help sail a boat, work on a dude ranch, drive racecars, or work on a farm. The attraction is not a product, not a thing, but an experience. In fact, some 47 percent of Americans have taken at least one adventure vacation in the last five years.

So, what do I mean by an experience? Because experiences are intangible, they are a bit hard to define. In truth, they don't have defining characteristics, but each experience shares a sort of family resemblance with other experiences. Here are some of the attributes they often share.

In an experience, services simply set the stage and products serve mainly as props

Joseph Pine and James Gilmore (1998) claim that experiences occur, in the commercial sense, when the services being delivered set the stage for the event, and the products being sold serve as props for the event, but the emotion elicited is the main affair.

Often, experiences are participatory

Experiences engage the customer in activities that require their attention. These activities often challenge the user to test his or her skill. REI offers a rain room where you can test the latest Gore-Tex® apparel. Usually you have to go to a waterpark to have this kind of experience. Nike Town offers sports clinics for their customers. Nowadays, I rarely go to a golf store which doesn't allow me to try out the putters and the drivers. Many golf stores have computerized golf simulators which enable you to play some of the most famous and challenging courses in the world.

Experiences are immersive

Experiences are often immersive, making the user feel like part of the production. Each instance of the experience is different because it involves different players. Immersion in an activity engages the user's attention and helps to make the experience memorable. Psychologists and educators have known for some time that learning by doing leads to better mastery than learning by observing. The doing makes the learning more memorable. This is also true of an experience. Getting people into the activity makes it more relevant and more fun.

Consider karaoke. What makes this form of entertainment so much fun and so interesting? First, it plays to people's fantasies. Without the customer, the show does not go on. Second, it is hard to forget (and still harder for your friends to forget!) an episode when you got up and tried to sing like James Brown. Third, it creates great stories to tell later.

Experiences elicit moods

This is what made my Nike Town experience so memorable. I felt inspired by it. I felt like anything was possible. The professions of usability and human factors have often neglected this aspect of design. You need to look to retail design, industrial design, and architecture to find people who really try to take this into account. People like this are often invaluable members of your design team. Now that we have some grasp on what an experience is, what principles can we derive to help us design better experiences?

Design experiences—not products

Forget about designing things. You are designing an experience, an event, a mood, a feeling. The products, the artifacts of your design, will influence the experience, but they should not be the main focus of your design. What do you want your user to feel? What emotions would you like to elicit from them? What do you want them to remember and tell others about the experience you designed?

Focusing on the entire experience gives you a more global perspective on the design criteria, and at the same time provides a more focused target for your design.

Design everything!

Every aspect of an experience can and should be designed. Don't stop with the knobs, dials, displays, buttons, and documentation. Design— and I mean really design—the packaging, setup procedures, customer service, troubleshooting, financing, method of disposal, and other aspects. All of these things directly affect the success of your customer's experience and consequently affect the future of your company.

Treat everyone as a designer

Since we are designing everything, it is important to think of everyone as a designer. I don't care if they're researchers, librarians, administrative assistants, users, or whomever. You never know where insights and design ideas will come from. Do not let ego or turf wars get in the way of great design. The lead designer will still have final say about how things get done, but why limit your options? Start looking to retail designers, marketing, and brand development strategists for interesting and insightful perspectives.

Take a new approach to user-requirements gathering

Focus on more than features and functions. Focus on more than task success rates, task completion times, errors, and requests for help. Focus on what experience the users would like to have, what emotions they want to experience, and how they want to remember your design.

Of course, we need to continue to follow the tenets of user-centered design (UCD): an early and constant focus on the user, deriving design guidance from human performance data and iteration. These principles apply to designing experiences as well as products. We simply need to extend them to more aspects of the design. This is a more holistic approach that requires new and innovative research techniques. We can discuss these techniques in future columns.

Reference

Pine, J., II and J. Gilmore, "Welcome to the Experience Economy." *Harvard Business Review,* July–August 1998.

Toward a Framework of Experience as It Relates to Interaction Design: Conference Workshop Report

Jodi Forlizzi,
Interaction Design

Over the past several years, we have witnessed a growing interest and enthusiasm for "designing the user experience." Our desire to conduct a workshop came in part from seeing these ideas spread to the mainstream business press (for example, the *Harvard Business Review*), as well as design firms' collateral materials. The real imperative, though, was that we felt that very little had been done to articulate the whole idea: What do we really mean by "designing the user experience" and how do we use interaction design and product design to achieve user experience goals? Without some firm grounding, we felt that "user experience" would become simply market-speak or a stand-in phrase for usability.

Our overall goals are lofty (to work toward a framework of how experience relates to interaction and product design). We saw the workshop as the beginning of a long conversation, one that could build a shared understanding and language and map out the areas needing further research and practice. In particular, we wanted to discuss:

- How can and should we talk about experience?

- What do we mean by experience?

- What are the types of experiences that users might want to have?

- What is the connection between product design attributes and these experiences?

Participants

Our workshop brought together experts from the United States,

Finland, and the Netherlands, with backgrounds in product design, business design, interaction design, user research, usability testing, arts and crafts, and theater and performing arts. The diversity of backgrounds enabled lively discourse and critical discussion.

Process

Before the workshop, we created a Web site to help participants become familiar with some of the models developed by other design researchers. We also asked participants to perform a self-documentation exercise, where, using a disposable camera and a log book, they recorded aspects of a particular experience. The purpose of this exercise was twofold: first, to get workshop participants to start thinking about everyday experience; second, to provide some initial discussion data.

During the workshop, we reviewed the self-documentation exercises, using them to aid in our discussions about different types of experience. We also generated examples from our collective memories. These examples were then used to derive interactions, product qualities, and qualities of experience, which were written on note cards and posted on walls in the work room. We also took some time during the workshop to discuss and analyze the existing models that we felt would be relevant to our process.

Finally, we attempted to look for patterns in the data by clustering our note cards and drawing out some useful continuums and stories of experience. Our work culminated in two building blocks for further exploration: defining what we mean by experience, and what connection product design attributes have to experience.

Ways to define experience

A difficulty we faced from the outset was that the word "experience" is general and ubiquitous. Much of our conversation revolved around ways to clarify its meaning while making it meaningful and useful to designers. On the whole, we felt that we made good progress on this front in the workshop.

While we are interested in how all types of artifacts affect experience (including products, services, and environments), we tried to limit our discussions during the workshop to user-product experiences. Because designers often interpret for an intended user in all points of an experience, understanding user-product experience seems like a logical place to begin, since designers need to understand how users will interact with a product. If designers can initially derive an understanding of these experiences, we may be in a better position to design not only product experiences but also experiences that enact change for the user, the product, or the context of use.

The simplest way to talk about experience is as the constant stream of thoughts and sensations that happens during conscious moments. This definition is inspired by cognitive scientist Richard Carlson's theory of consciousness known as "experienced cognition." Self-talk or self-narration is often the way that people acknowledge the passing of this kind of experience. This seems to be the level of experience that designers refer to when talking about stimuli such as visual design elements or about task flows and transactions.

Another type of experience, known as an experience, is described by philosopher John Dewey. This type of experience has a beginning and an end, and changes the user, and often, the context of the experience, as a result. A simple example of an experience is a fine meal in a restaurant which comes to define or epitomize a fine dining experience. People often acknowledge this type of experience by telling stories about the experience, or by using the experience to compare and judge new experiences. When designers wish to provide a particularly memorable or category-defining experience, they may be referring to "an experience."

The connection between storytelling, memory, and experience is an interesting one, and has been discussed at length by Roger Schank, a cognitive scientist. Stories are the ways that people condense and remember experiences, how they choose important threads to communicate in certain situations, and how experiences are relayed from person to person. Experience as story plays an important role in events as diverse as courtroom proceedings and television soap operas. Designers could think about a user experience in terms of what kinds of stories we want users to tell about our products, as well as what the gist of the user-product experience might be. A positive gist could help build long-term brand loyalty; for example, positive stories could support word-of-mouth advertising.

Workshop participants agreed that these ways of talking about experience are related. For example, an experience can be made up of an infinite number of smaller experiences, relating to other people, settings, and products. For example, in a story of a lunchtime walk, one participant talked about the constituent experiences of reading a book, looking at a favorite tree, tripping on a step, and having a chance conversation with an old friend. Likewise, the stream of sensations experienced during a fine meal will affect whether that becomes "an experience" and if it does, the qualities it has for the user.

In addition, experiences have different amplitude (some play heavily on our emotions, and some are simply about accomplishing a task with minimal effort). Some experiences we want to collect (for example, seeing a movie or visiting a painting in an art museum), and some we don't want to repeat (travel delays or difficult interactions with nasty clerks). The group did make a point that not all intended experiences had to be positive; for example, aversion training to quit smoking works on the principle of creating a bad experience.

Finally, we felt it important to make a distinction between designing experience and designing for experience. Experience is personal, and we must realize that there are wildcards or factors that we can never control by our designs: personal interpretations of a situation, based on different cultural background or prior experience; emotionally aroused states which cause different subjective interpretations of a certain moment; and the element of chance, when events coincide in a random way that seems meaningful to a certain person. For example, one participant recounted a story in which a cell phone allowed a colleague to make an important business connection without breaking the rhythm of the party he was engaged in. The value of the cell phone became escalated as a result of this event, which had happened by chance. This is not to say that it is impossible to support intended experiences; if this were the case, Disney would be out of business. What it does mean is that as designers thinking about experience, we can design only situations (levers that people can interact with) rather than outcomes that can be many exciting arenas of research and discovery.

To understand how designers might define the ideal experience, we need methods for understanding what kind of experience is appropriate for a particular audience in a particular context. We adapt various qualitative research techniques, such as ethnography and contextual inquiry, to support this area. We believe that making distinctions about types of experience, as described in above, will also help designers define meaningful targets. For example, during our discussion, we talked about how the starting and ending points of a story could be broadened to create better environments for user experience. An airline thinking about improving the flying experience might want to think beyond what happens while the passenger is in a seat on the plane to include what happens in making travel arrangements, to traveling to the airport, to spending time at the gate.

A great deal of work needs to be done to understand how to connect a product design to a specific experience. A building block will be to better understand the principles of how people interact with various artifacts and how user-product interactions affect the experiences people have. During the workshop, we generated a number of examples, some of which were conveniently supplied by the conference itself and its venue, the ultra-luxurious Phoenician. For example, the conference is small, intimate, and collegial, which is supported by its design: The location is a semi-isolated resort, conferees eat almost every meal together, name badges prominently display the wearers' first names only, preventing the worst kind of badge-sniffing.

On Thursday night, Brenda Laurel gave a keynote address to the conference during an outdoor dinner in 112° heat. During the talk, the wait staff brought around ice-cold washcloths—a design detail that was appreciated and memorable. The Phoenician was also full of examples of how expectations color our experiences (from the large ostentatious bathrooms to the somehow inauthentic Native American dances at the Thursday night dinner).

Although we tried to cluster the data from these and other examples to create some sort of diagrammatic representations to support design thinking, we came away realizing that it may not be possible to create a single model for every situation. We also revisited the existing models to see if they would help us muddle through, but found that they had their own problems. Clearly, there is much more work to be done in this area.

Conclusion

As with most workshops, probably the best thing we took away was the chance to spend several days with some really bright and wonderful people. We hope to continue working with them, hearing about their individual projects, and finding new colleagues to join our conversation. We thank the conference organizers for giving us the opportunity to discuss this exciting topic.

For more information, please see

http://www.goodgestreet.com/experience.

Jodi Forlizzi
Interaction Design
jodi@goodgestreet.com
http://www.goodgestreet.com

Ease of Learning and Efficiency of Use: Usability Needs to Focus on Both

Russ Branaghan, Fitch Inc.

Usability is complex. Now that may sound like a strange thing to say, but usability really *is* complex. Usability is not a singular entity but a combination of several qualities rolled into one. Nielsen (1993) lists five components of usability: ease of learning, efficiency of use, memorability, error avoidance and handling, and satisfaction.

Usability engineers are often criticized because they focus primarily on ease of learning while quietly sweeping the other four qualities under the rug. For example, programmers often point out that a usability test may be testing only ease of learning and that the usability engineers have no idea how efficient the user will be after having worked with the system for some time. Of course, they are right. Many of us often do avoid the important issue of efficiency of use.

Ease of learning

Ease of learning is important for a number of reasons:

1. The user's initial experience with the product means learning to use it.

2. Trade press reviews often equate ease of learning with usability. They rarely test the system long enough to judge its efficiency of use.

3. People may use lots of technology products, but they become truly efficient with only a few. They need products to be easy to learn.

Ease of learning is particularly important for products that are used only once or infrequently, such as some kiosks and Web sites. Ease of learning is also important for business and personal productivity applications. With these products, the user does not employ all of a product's functionality but needs to learn some of the functionality quickly.

There are several tried-and-true methods for making products easier to learn. As usability engineers, we focus on good organization of the content, an appropriate and consistent metaphor, direct manipulation, intelligent defaults, and recognition rather than recall. Figure 1 shows a kiosk that Fitch designed for Chrysler at the Mall of America in Bloomington, Minnesota. It uses these methods to ensure that users can simply walk up and use the product. The users, tasks, and environment persuaded Fitch to focus on ease of learning rather than ease of use.

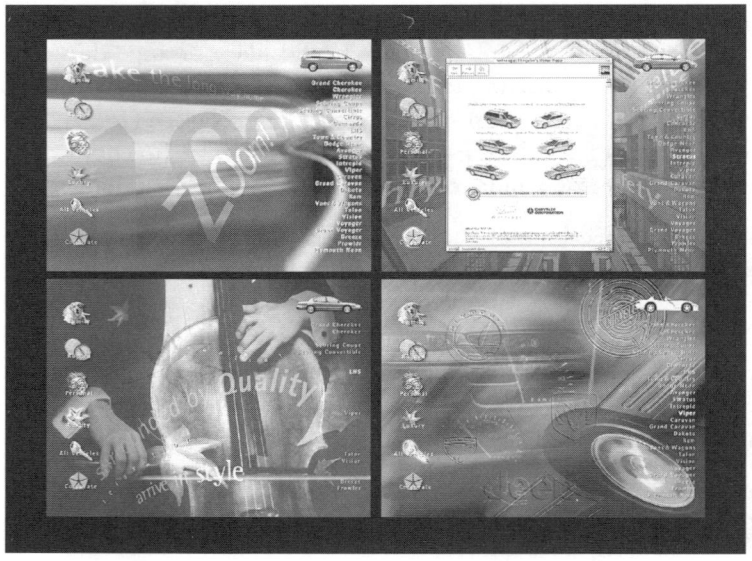

Figure 1. This interactive kiosk for the Chrysler Store at the Mall of America must be very easy to learn.

Usability engineers often test ease of learning because it is easiest to test. All you need to do is to find novice users and ask them to do some realistic tasks with your product or with a prototype of your product. The tests are often short in duration, require no training of the users, and incorporate some fairly simple performance measures such as success rate, time on task, time to reach a predetermined level of performance, and time for performance to plateau.

Efficiency of use

Efficiency of use refers to the users' productivity once they have made it close to the top of the learning curve. Efficiency of use may be more important than ease of learning for users who use a product frequently, who are provided with training, and who are motivated to learn.

Efficiency is improved by using accelerators, type-ahead, click-ahead, right-mouse menus, default value entries, and so on. An example of a product which focuses on efficiency of use is shown in figure 2. This product was designed for an industrial control environment. The product's users employ the software daily for control, monitoring, and troubleshooting. Although it was important that it be easy to learn, it was much more important that it be efficient to use. Efficiency facilitates the diagnosis and correction of problems.

Unfortunately, testing products for efficiency of use is more complicated than testing for ease of learning. To begin, you need to find and qualify experienced users. You can find such users if your product has already been released or is in use at beta test sites.

On the other hand, if your product has not been released, you must train novice users on your product until they are well along the learning curve. One way to ensure sufficient skill is to train users until they no longer show improvement. At that point, you may assume that the users' performance has reached a plateau and that they have reached a level of skilled performance on that product.

Testing a product's efficiency is often difficult, if not impossible, to do with systems which are in early design or development. There may not be enough of the system implemented for the users to really develop any skill with it.

A combination of goals

In most cases, we want to design a product that is both easy to learn and efficient to use. This combination of goals is one of the most significant challenges for designers. And often, one design may satisfy both criteria. A system that is easy to learn is often efficient to use because it is well designed, functional, and consistent. Figure 3 shows a user interface for the Texas Instruments Avigo Personal Digital Assistant (PDA). In this case, the users needed to be able to quickly learn the simplest features such as the calendar and address book, but also needed to be quick and

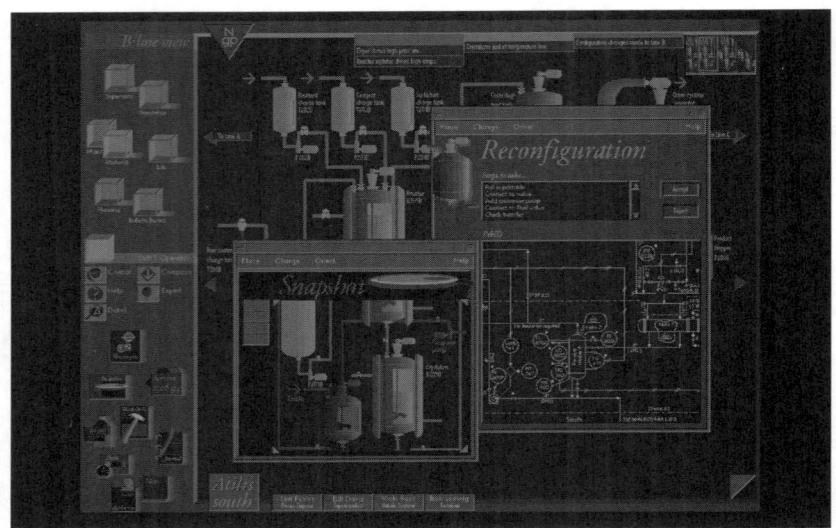

Figure 2. A design for an industrial control environment emphasizes ease of use.

efficient once they learned to use the product. Employing both types of usability testing (ease of learning and efficiency of use) ensured that these goals were met.

As usability engineers, our responsibilities do not stop when the learning curve reaches a plateau. Our job is to design for usability, and our responsibilities extend to the accomplished expert as well as to the novice. We cannot neglect that responsibility by conducting only the simplest usability tests.

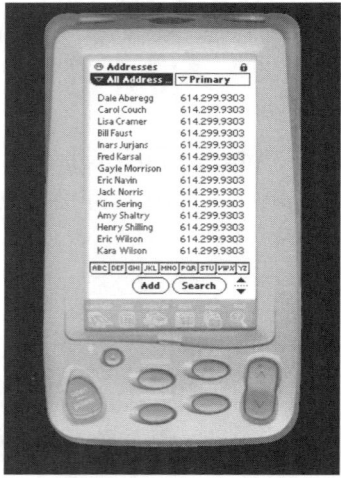

Figure 3. The PDA interface supports both ease of learning and ease of use.

Reference

Nielson, 1993 reference.

Consulting Issues

Authentic Consulting: Bringing the Power of Who You Are to What You Do

Jeff J. Rubin,
The Usability Group
(TUG)

Introduction

With technical skills responsible for only a small portion of the success on a project, it is vital that we, as usability professionals, develop our ability to interact authentically and skillfully with clients and coworkers. At the same time, we face unique challenges having to do with lack of appreciation, indifferent or hostile colleagues, and resistance to using our expertise effectively. Authentic consulting involves being true to our beliefs and values while facilitating and provoking change in the organization. It addresses both the psychological and logistical issues of internal and external consulting.

What do we mean by "authentic"?

Authenticity or genuineness is hard to define, and difficult to recognize in one's own experience. It has to do with the depth of one's personal connection to one's being, and how one behaves and relates with others under a variety of conditions. From my own experience and study, authentic behavior is characterized by the following:

One's speech and actions are synchronized

That is, what we say and what we do are integrated and in tune with each other. Both our verbal and non-verbal messages are consistent, and we are not trying to hide how we feel in order to manipulate situations.

Vulnerability, lack of arrogance

Arrogance is a form of one-upmanship—trying to come out on top by putting others down. Since it is based on a subtle form of insecurity, it is difficult to work with an arrogant person because opposite viewpoints are often seen as a threat. Authenticity, conversely, is evidenced by a sense of vulnerability and accessibility. Such people are approachable, reasonable, and willing to hear viewpoints different from their own.

Consistent demeanor

How many of us feel, from time to time, that we lead two or more lives? Or feel that we are being pulled in vastly different directions by our work and other interests and obligations? For example, one persona at work, one persona at home. While authentic behavior is appropriate to a particular situation, it is also evidenced by a consistent demeanor overall. One does not act one way in the office with one's colleagues, then go home and "kick the dog." There is an evenness to one's behavior, and one is not constantly roiled by hope and fear.

Aliveness, vitality

Lastly, authenticity means being in touch with the simplicity of being alive and in the moment, and the ability to do one thing at a time properly and wholeheartedly. In fact, if there is one key element to being genuine, it is a sense of being 100 percent "here" and not neurotically involved in fear of the past, and hope for the future. Such a person will appear to be grounded in reality, and that reality is always "now."

What are the obstacles to remaining authentic?

As romantic and inviting as the concept itself might sound, it is not easy being and remaining authentic. In fact, most of the time, we might notice just the opposite, that we are behaving in ways that are at odds with our true intention. Some of the obstacles to behaving authentically are:

"Going along" behavior

We know something does not sit right with us, such as a major decision on a project, but we go along with the majority opinion, because it's expected of us. For example, our boss just asked our entire group for their input on his new idea, and we think the idea has little chance for success. We have a bad feeling in our gut, but instead of voicing our opinion, we decide not to make waves, and we tell him it is a fantastic idea.

Ignoring one's own needs and conviction

Another obstacle to genuineness is when we sublimate or rationalize our own needs in the interest of someone else's. For example, when contracting with a new client, we know we need management backing for a new process we are instituting, but we neglect to mention that at the outset since it is not what management wants to hear. Instead, we ignore our instincts and plow ahead, knowing full well that the project has a small chance of ultimate success.

Conflicting emotions

We experience a lack of focus, and a sense of being caught between wanting something and not wanting it. You could say we are not sure whether we are coming or going. On one hand, we feel very positive about our situation, and our lot in life. On the other hand, we are not sure a major change might be appropriate.

Blind spot

The classic expression "what problem?" aptly sums up this last obstacle. We are so far out of touch with ourselves, we do not even notice anything is "off," or that we are acting in a conditioned manner. If people challenge us about our state of mind, we are defensive and pretend not to know what they are talking about.

An important thing to remember about these obstacles is that they are not "solid" or fixed, in the sense of being inherent parts of our personality. Rather, they are temporary experiences which tend to become habitual ways of behaving over time. The power of the obstacles (to authenticity) comes from our lack of acknowledging what is happening with us, from not paying attention. Yet the obstacles have one thing in common, which is the key toward overcoming them: All the obstacles are based upon fear.

Working with fear

There is a three-fold logic in working with fear. I call it the three R's:

Recognize (fear)

Relate (with fear)

Results (of confronting fear)

Fear is expressed by anxiety, insecurity, nervousness, feeling inadequate, loss of confidence, and depression. It arises when we are challenged, or when we are venturing out of our "comfort zone." But fear is not the problem. The real problem is not recognizing and acknowledging fear, for there is a habitual tendency to cover up fear and ignore it, to "push it under the rug." This is unfortunate, since from the viewpoint of behaving authentically, fear is "good news." It indicates that we are getting to the heart of the matter on an issue of importance to us, and is the natural result of growth and extending ourselves. It is the vanguard that something is resonating within us at a deeper level beyond intellectual niceties. Fear contains tremendous intelligence, since it is an aspect of our intuition about what is really happening with a situation. When we repress fear, we lose touch with a vital aspect of ourselves.

Relate (with fear)

The second step has to do with how we relate to fear. Unfortunately, we've been taught that fearlessness is the absence of fear (the "James Bond school of warriorship" if you will), which is why we tend to repress, ignore, or simply steamroll our way through fearful situations. In actuality, fearlessness is possible only in the midst of fear. Thus, by ignoring fear, there is really no way to work with it or see its value.

The problem is our belief that fear is a sign of weakness. In contrast to ignoring, the skillful way of working with fear is to welcome it when it arises, and to actually go toward it. For example, suppose we become fearful during a meeting devoted to presenting our findings of a controversial study. The study found that the product was poorly designed, and we know that this will not go over well with the audience. Normally, we might "sugarcoat" the results to avoid confrontation, but in this case, we just decide to tell the truth and see what happens—even if it means breaking into a cold sweat and stammering a bit. From this point of view, relating with fear is a stepping stone toward greater awareness and toward a more skillful and creative way of working with ourselves and our clients and colleagues.

Relating is not simply an intellectual exercise. It evolves from an intense desire and commitment to be who we are. Mindfulness practice, such as meditation or yoga, can help immensely. Through practicing, we allow those "undesirable" aspects of ourselves, including the fear of being exposed, to shine through to the surface where we can acknowledge and work with them. However, it is important to remember that this is a long-term project rather than simply a one-shot deal. We're not trying to get rid of fear, but to transform it.

Result (of confronting fear)

The result of our willingness to confront fear is a breakthrough into a different way of relating to situations. It is the source of "outside the box" thinking because we are going beyond our habitual ruts. Much of our energy that typically goes into repressing and ignoring our instincts is now freed up for creative process. Consequently, insight arises, however shaky that might be at first. The three R's are a method of working with the obstacles toward being ourselves and go hand-in-hand with authentic consulting.

Authentic consulting: the contracting phase

If authenticity has to do with fearlessly being true to ourselves, how do we apply this to our consulting work? The definition of consulting, according to Peter Block, is whenever "you try to change or improve a situation, but have no direct control of the implementation." In other words, authentic consulting has to do with being true to our own beliefs while provoking and facilitating change in the organization. We are trying to influence others to do the right thing as we experience it, and the key moment for influencing is right at the beginning, during the initial contracting phase. It is the time when we assess what the client, boss, or colleague wants to happen on a project, and what he or she wants from us.

Usually what the clients want is clear. They want a study or a usability test conducted, or some research performed. But what do *we* want? This is the aspect of consulting that we gloss over, as if we did not have specific needs on a project. It is the point at which fear arises, and is often the point when we rationalize away our own needs on a project. When Peter Block polled some internal consultants, here are their most frequent responses about what they wanted to happen on a project:

• Clear definition of the job

• Work on the problem together

• Commitment to the project

- Sharing the blame and the glory
- Openness to feedback about the project
- Feedback on what happened after I left
- Feedback to my boss

Note that these are non-technical issues, and many of them have to do with improved communication. The time to address these issues is the contracting meeting. After that point, it is difficult to go back and revisit these issues, although still possible. During the contracting meeting, it is important to ask for what you want simply and directly, rather than getting into long, theoretical rationalizations.

Asking for what you want: four steps

Let's assume that you have been charged with educating a group of developers about user-centered design, and you sense that they will be highly resistant. You feel it is absolutely critical for their manager to also attend the course to set the tone, and show management support, even though the manager doesn't want to take the time. Here are four steps to acting on what you want:

1. Put what you want into words, simply and directly.

First, a poor example:

"We have found from research in the area of instructional technology that learning in a training course tends to be retained and used more efficiently if there is tangible evidence of positive reinforcement back on the job. This allows for better use of the newly acquired skills and, in your case, the user-centered design course is introducing a totally different approach to the way employees perform their jobs and view their work. If this positive reinforcement does not take place, then you can expect a degradation of information and the cost/benefit algorithm of your investment is severely decreased. Have you ever attended a course on user-centered design?"

Notice how this example is indirect, non-assertive, and lets the manager off the hook. Then, a much better example, as concise and direct as possible:

"I would like you to attend the course on user-centered design."

2. Let the client/boss react.

Example: Oh?

3. If you get a question, give a simple two-sentence explanation, restating what you want.

"Yes, we've found that training is much more effective when management shows their commitment and interest by attending, as well as sharing their unique perspective during class discussions. I'd really like you to attend."

4. Wait for yes or no.

This is often the most difficult part, since you'd like to fortify your case. But silence often communicates determination, and a willingness to stick to your guns. If the answer is still "no," and you go forward with the project, it is important to go on record with your rejected request. It can be used as a negotiating point in the future.

Furthermore, it is important to put what you want in writing. Internal consultants can draft a memo outlining their understanding of their responsibilities and the different roles on the project. External consultants can include a "client responsibilities" section that includes the issues of importance. For example, the client agrees to:

• provide access to certain documents and people

• identify equipment to be provided

• distribute results to the entire team and upper management

• coordinate attendance of observers for a usability test.

Consultant as change agent

If consulting involves provoking change in an organization, then it's important to discuss the role of consultant as change agent (versus the researcher). Consultants acting as change agents will design a study so it focuses on what to do next, on an action plan.

A study should address a specific problem rather than just being a collection of data, the eventual use of which is unclear. In my experience, I have seen countless studies conducted that simply collect dust or that never get distributed to the correct audience. Eventually, the collection of information becomes the status quo, as in, "We always do focus groups to establish our initial market," or, "We always conduct a usability test every six to eight weeks." However, if the information isn't used to change the product or the organization, why do it?

Use plain language so that the transfer of information is enhanced.

Don't fall into the trap of trying to impress by using overly technical terms. As usability specialists, we are especially vulnerable to this seduction because the profession is often undervalued and maligned. Instead, use language that the average person can understand. It will help immensely in establishing a collaborative relationship.

Include the "client" in deciding how to proceed, and handle resistance as it occurs.

Use a collaborative style of consulting, rather than acting as an expert. While your technical expertise is unique and valuable, it is absolutely critical that the client share responsibility for implementing the results. While it might be tempting to take on full responsibility for deciding what to implement, especially if you are seen as the expert, it is a temporary victory. Consultants simply do not have the appropriate standing in the organization to implement fundamental change by themselves. Invariably, things hit a snag or do not work as expected, and clients quickly become disgruntled. Should your clients resist collaborating on how to proceed, either by "not having time to meet," or flattering you by saying, "You know best, you've done this so many times," I encourage you to reason with them and explain the need for their participation. Otherwise, your chances for long-term success are practically nil.

Address how the organization functions both in the problem statement and your recommendations.

Projects invariably are composed of 10 percent technical issues and 90 percent human and communication issues. The disregard for this truth within organizations is why so many technology projects fail, and why the cartoon "Dilbert" strikes such a chord in all of us. Therefore, when conducting your initial analysis, study how the organization works, and learn its culture. Who holds the power position—marketing or R&D? How are developers viewed within the organization? How willing are they to implement new approaches? Why were you brought in? It makes no sense to recommend conducting contextual inquiries at a customer site to organizations that have never even conducted a usability test. It is simply too great a leap. By asking the right questions of the right people, you can quickly ascertain the organization's openness and resistance to change, and design a study accordingly.

Consultant as leader:
how do you know when you are ready to lead?

Provoking change invariably requires taking the lead on a project from time to time. How do you know when you are ready to lead? You are a candidate to lead when you hold strong beliefs about your work and envision more ideal ways of working, both for yourself and the organization:

- you have strong opinions about which way your clients and colleagues should move, and a sense of frustration at how things are today

- you want to have an influence and are action-oriented.

In leading project teams, begin by clarifying the difference between what exists today, and what the team wants. This involves helping teams formulate what they want to accomplish, and expressing it in terms of an overall concept or vision. Once a vision for the project is formulated, bring new perspectives and approaches and encourage others to feel they can make a difference. However, don't simply become a cheerleader with your head in the clouds. The crucial point is not to get so completely lost in the vision that you ignore the severe problems of the organization. It is essential to stay grounded, to stay in touch with the real problems without being overwhelmed by them. You need to find a balance between vision and practicality. Too much reliance on vision brings wishful thinking and impractical solutions. Too much reliance on practicality brings timidity and remaining with the status quo.

Conclusion

Authentic consulting begins with a willingness to face fear and acknowledge that we, too, have needs on a project that speak to our heart, and our ability to act in creative and novel directions. By facing fear, we tune into the intuitive, gut-level aspect of seeing, rather than simply manipulating intellectual or analytical concepts. We discussed the contracting phase as the initial way of addressing our needs as well as our clients. By practicing authentic consulting through the life cycle, not only do we experience a more satisfying way of integrating our work with our lives, but our clients benefit by getting much more of what we have to offer.

References

Bellman, Geoffrey M. (1990). *The Consultant's Calling.* San Francisco: Jossey-Bass.

Block, Peter (1981). *Faultless Consulting.* Austin, TX: Learning Concepts.

Senge, Peter M. (1990). *The Fifth Discipline: The Art and Practice of The Learning Organization.* New York: Doubleday.

Trungpa, Chogyamm (1984). *Shambhala: The Sacred Path of the Warrior.* Boston: Shambhala Publication, 1984.

Consulting Skills for Usability Professionals

David Gilmore, IDEO
Product Development,
and Derek Millard,
Enable Change, Inc.

At the 7th Annual UPA Conference (Washington, D.C., June 1998) about 40 participants attended a day-long workshop in "Consulting Skills for Usability Professionals: Enhancing Successful Client Relationships." As part of the initial exercise, participants identified what they personally most wanted from the session. Then the group as a whole summarized these learning goals into four themes:

- How to be more effective and improve impact

- How to get more business (external consulting) or better utilization (internal consulting)

- How to deal with difficult clients

- Tips and techniques in general.

Rather than try to cover all of the workshop content, we have summarized the key points for each of these themes.

How to be more effective and improve impact

Although a usability professional's expertise is in the field of human factors or human-computer interaction, the process of delivering that expertise to a client is called consulting. Effectiveness is a property of both the technical expertise and the skills of delivery—it isn't "how much you know," but "how much of what you know is used."

Effective use of your expertise is about impacting important business outcomes for your client. Improving your effectiveness can depend less upon your technical expertise than upon your consulting skills and process of delivering that expertise.

Be more conscious of the consulting process

Consulting is a process of delivering your expertise to a client where you do not have direct control. The five-stage process occurs through a series of meetings, conversations, and interventions. In effect, the process is a loop—but one that need not be completed before it begins again (figure 1).

To help increase your awareness of the consulting process, consciously ask yourself, Where in the loop are we now? and, Where do we want to go next? At each stage ask, What are the task issues involved? and, What relationship issues are likely? You will add more value when you can pay attention to both task and relationship issues at each of the five stages of the consulting process.

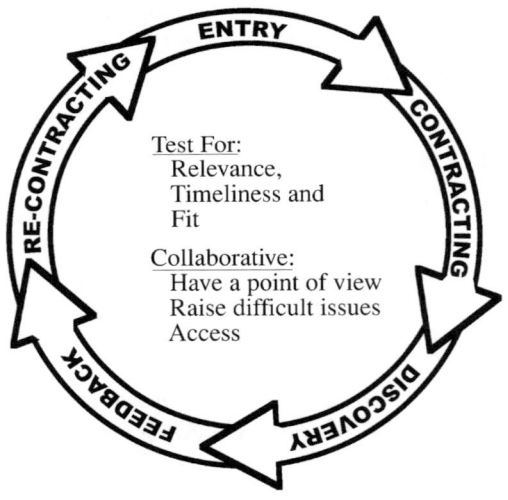

Figure 1: The five-stage consulting process.

Key issues to pay attention to throughout the consulting process:

- Who is the client?

- Is there relevance, fit, and timeliness?

- Is the relationship collaborative?

Who is the client?

Sometimes it is obvious who the client is, but it can also be ambiguous. It is not the definitive answer which is useful, so much as the clarity that can come from exploring the question.

You may well have "bankers" (who are funding the project), "bosses," major stakeholders (end users, customers, developers, etc.), as well as the actual client. You might have multiple clients, or you could be consulting to a consultant who is, in turn, consulting with an eventual client. Another version of this question is to ask yourself, Where is my expertise going to be used? or, Where am I going to add value?

Not only do you need a firm grip on who the client is, but you also have to be clear about the nature of the agreement or contract with that client—mainly what you are doing together and how you are going to proceed with each other. Your agreements with the client need to be in accord with any agreements made with other stakeholders. Failing to deliver on your agreements will put your credibility at risk.

For example, in negotiating access to an end user population for usability testing, you need to know what you are capable of delivering— gaining access by promising them the software of their dreams may be an effective strategy in the short term, but it can turn into a disaster when you are unable to deliver on your promise.

Are there relevance, fit, and timeliness (RF&T)?

Will the application of your expertise positively impact the client's business outcomes—is your expertise relevant?

Does your approach fit with the client's way of doing business or with the way they wish to use your expertise?

Is your intervention timely—do you and the client have sufficient time, energy, and resources to implement your methods?

Relevance, timeliness, and fit are all essential for the success of the consulting process in terms of adding real value to the client. They can be thought of as prerequisites for success. The client may need to make changes to achieve fit and timeliness, even when your usability expertise is obviously relevant.

It is unusual to know in advance to what degree there is RF&T. Indeed, it is often a good idea to make your first consulting contract with the client an assessment of RF&T. Then, consult with the client about making changes needed to achieve the necessary fit and timeliness. Remember—the goal is for your expertise to have a positive impact on business outcomes. Without RF&T, that won't be possible.

Is the relationship collaborative?

There are three approaches to the client-consultant relationship. The consultant can be a "pair of hands," an "expert," or a "collaborator."

Each of these approaches has its own particular costs and benefits—both to the client and to the consultant. Which type of relationship you are in is derived from who identifies the problem and who identifies the solution. If the client identifies both, the consultant is likely to be functioning as a "pair of hands." If, on the other hand, the consultant identifies both, the consultant has an "expert" role. Finally, if the client and the consultant together identify both, then the relationship is *collaborative.*

A particularly hazardous situation is one in which the client identifies the problem and wants the consultant to identify the solution. The client is asking you to be a pair of hands and an expert at the same time. This confusion over the nature of the relationship makes it easy to blame the consultant when outcomes are less than hoped for.

If the desire is to impact the client in a way the client values and for your expertise to be well used, then a collaborative relationship is needed.

To establish the exact nature of the relationship is one of the goals of contracting—when consultant and client talk about their wants. To create a working collaborative relationship, the consultant has three essential wants:

- To have and be able to express an independent point of view (based on your expertise)

- To raise difficult issues with the client

- To have access to the client, to information, and to the organization.

From the point of view of your expertise, you can't test for relevance, timeliness, and fit without agreement to your consultant wants. And without RF&T, the business risks wasting resources and failed implementations. Be prepared to make this business case to your client, otherwise they might not agree to your wants.

Of course the client may not want a collaborator—or you may prefer to be in an expert or pair-of-hands role in a particular situation. What is critical is that you and the client, at any point in time, have a shared understanding of the nature of your consulting role. If you have very different role expectations, there are bound to be difficulties.

Getting more business and your expertise better used

The key to getting more business or improving the use of you and your skills is to build credibility within the organization and with your potential clients. Obviously, past successes are a strong builder of credibility—providing your expertise was clearly understood, and providing the client connects your contribution with improved business outcomes. Until you can establish that track record (or if you need to overcome a past bad one), you must clearly articulate your expertise in terms the client appreciates and make an offer connecting successful use of your expertise with improved business outcomes. For example, will using your expertise:

- Help provide the client with a competitive advantage in some way you can define?

- Help the client provide a unique response in their marketplace?

- Improve the client's capacity to concurrently increase profitability, improve quality (of product and/or service), and reduce cycle time?

Your ability to logically connect your expertise to these outcomes is a powerful way to build credibility. If you feel undervalued or that your expertise is not well understood or well used, you're probably right! Take the initiative. These are difficult questions to address, but well worth the time and effort as you begin to have more meaningful conversations with clients about what you have to offer and how it can help them.

Dealing with difficult clients

One key is to stop seeing them as difficult! They are just being clients! Seriously, if all your clients seem difficult, perhaps you are somehow contributing to the problem. If a client seems much more difficult than the norm, initiate with good faith and goodwill a contracting conversation with them. Summarize what goals you share, how you would like to work together differently, and what you are willing to do. Ask them what they would like from you and tell them what you need from them. This has the potential of renegotiating your relationship and beginning again on a new, sounder basis.

However, even with the best of clients, there will be difficult moments—you can expect them. In fact, some difficult moments are sure to arise if you are dealing with real issues. Getting both you and your client through these difficult moments is part of your job as a consultant—don't expect the client to assume that role. Instead, the client will simply distance themselves from you or not use you next time. It is your task to keep cool, figure out what is going on, and know what to do about it.

Fundamental sources of difficulty

There are three fundamental sources of difficulties with a client:

1. The client does not see you as having anything of substance to offer. If this is the case, managing the relationship is not going to help. If you suspect this is the situation, deal with it directly by doing your homework (see "Getting More Business and Your Expertise Better Used" above). Then, in a calm and non-defensive way, raise the issue, describe your expertise, and make an offer. If the client is being forced to use you—for example, if a third party has mandated the project, then raise that issue in a neutral way with the client. Don't take sides or use the mandate for influence. Instead, talk about your expertise, and the need to explore relevance, timeliness, and fit. See if you and the client can set aside the barriers created by the mandate and proceed on a businesslike basis with each other.

2. If you and the client have different expectations of your consulting role, you are in for difficulties galore! For example, if you are in collaborative mode and the client sees you as a pair of hands or expert, then neither of you will have your expectations met. The only way to get to shared expectations is to talk about them, and when things seem to get off-course, talk about them again.

3. The client and you are in some form of power struggle. Do you see the client as an adversary? Someone to overcome in order to be

successful? When your client takes a position, do you jump to arguing for your point of view without really listening? Any of these are clues that you are not consulting; at that moment, you are doing righteous battle. And most likely you will lose . . . you are the consultant. All our relationships involve power. It is human to want to win or at times be taken care of. However, in the consulting relationship, we must choose to reach instead for that part of us that values others and wants to make a contribution. If you find yourself in a power struggle with your client—PAUSE. Is this really where you want to be? Are you willing to let go of it on your part? It is an inside job—managing yourself can be a difficult part of consulting.

Dealing with resistance

Another type of perceived difficulty is resistance. Resistance is not the same as disagreement. If the client openly disagrees with you, or candidly tells you their objections and concerns, deal directly with the disagreement by making your case, in business terms, for your point of view and hearing their case. See if you can't reach agreement, or at least understanding. Realize that the client has the final call and the final accountability. Resistance is an indirect expression of concern usually around control or vulnerability. It can take many forms:

- silence

- confusion

- getting lost in details

- changing the subject

- superficial compliance

- ridicule.

Resistance is seldom conscious and always means something real is going on. In that sense it can be seen as a good thing, rather than an enemy. However, if you don't deal with it, it will only get worse, making it impossible to proceed with meaningful content.

A sound preventative is to frequently ask your client what they think—whether they have questions, concerns, or reservations. If you can get a direct expression of these, then you won't have to deal with the indirect resistance. But, if a pattern of resistance continues, then it has to be dealt with before you can return to content.

First, hold on to your goodwill toward the client. This can be very difficult, as you will feel a tug toward your own emotional response (usually related to control, approval, or competence). So, if you find yourself pushing harder or searching for acceptance—STOP! Then, with goodwill, say in a neutral way what you perceive to be happening and shift responsibility toward the client. You can do this by:

- asking for reservations
- asking them what they want
- asking for what you want.

Remember that client resistance usually arises from unstated concerns about control and vulnerability. By saying what you see and shifting responsibility to the client, you are returning some control to them. Your goal is to get a direct expression of the client's concern. Then, together, discuss how it can be resolved.

Tips and techniques

Consulting is a complex, multidimensional process. There is no real recipe that will apply in all situations; however, there are a few additional tips that have proved generally useful.

- See the client as your friend. This doesn't mean acting in a "back-slapping" manner; rather, internally reassure yourself "this is my friend." Remembering this in a difficult moment can dramatically shift how you feel, enabling your goodwill to come through, and thereby change the outcome.

- Get yourself out of the way. If you find that you are taking things personally, or that you are fighting the client, or seeking acceptance/reassurance, then it is time to remind yourself that this isn't about you! At least, not at that moment—pause, take ownership of your contribution, and move on. Later, maybe call a friend or a shadow consultant and let loose, processing your feelings.

- Be clear about your own intentions. Think about your general intentions as a consultant. Are they to add value, to create commitment, to be fully present? Reach for your explicit intentions to help carry you through difficult moments. It is also helpful to be clear about your specific intentions for a particular meeting or conversation. What am I trying to accomplish? Where are we trying to go?

In summary and learning more

Being an effective consultant is about delivering your expertise in a manner that uses your skills well and positively impacts the client's business outcomes. Your overall effectiveness may depend as much on your consulting skills as on your expertise itself. By being more conscious of the consulting process; raising issues around relevance, fit, and timeliness; being able to clearly articulate your offer; and dealing with resistance in a positive way, you can improve how your expertise is used and add more value to your client's business.

Consulting is about continuously learning; to get better at consulting requires practice and consciously paying attention in the process.

Developing strong consulting skills takes more than is possible in a short article. There are a number of ways to learn more about consulting and the consulting process. The following resource list includes some of the most useful books on the subject, available publications, and sources for various types and lengths of consulting skills workshops. Other options include retaining a formal shadow consultant (your consultant on consulting) or agreeing with a trusted colleague to use each other as shadow consultants. Another effective strategy is to create a consulting "clinic" by establishing a regular time to meet with a group and focus on the consulting process, client situations, issues, and learnings.

Resource List

Books

Joel P. Henning. *The Future of Staff Groups: Daring to Distribute Power and Capacity.* Berrett-Koehler Publishers, 1997. Focuses on adding value to the business and articulating your offer.

Peter Block. *Flawless Consulting: A Guide to Getting Your Expertise Used.* Pfeiffer and Co., 1981. Focuses on the consulting process and relationship issues including resistance.

Lesley Trenner and Joanna Bawa. *The Politics of Usability.* Springer-

Verlag, 1988. Includes a discussion of some of the more technical usability issues in client relationships.

Publications

Marshall House Journal—Personal empowerment and transformation mostly from a consulting perspective. Published monthly. For free sample copy call 310-458-1172, or e-mail jeanie@mhmail.com.

Consulting Today—Practicing Consulting in Today's World. Published quarterly. For free sample copy call 914-591-5522, or e-mail consul2day@aol.com.

Consulting Skills Workshops

Enable Change, Inc. Custom in-house workshops of any length, facilitated consulting clinics, and individualized shadow consulting. Contact Derek Millard at 415-337-5045, or e-mail derek@enablechange.com. Designed Learning, 3- to 4-day public and in-house workshops. Call 908-889-0300 for catalog.

The NTL Institute for Applied Behavioral Science. 4- to 5-day workshops. Call 703-548-1500 for catalog.

The Gestalt Institute of Cleveland. 4- to 5-day workshops. Call 216-421-0468 for catalog.

Acknowledgment

We gratefully acknowledge that Joel Henning and Peter Block originally expressed many of the concepts and ideas in this article. Their books are included in the Resource List.

It's Time for a Usability Code of Ethical Conduct

Don Ballman,
Human Factors Architect

Welcome to a new column which will appear periodically in *Common Ground*. My objective here will be to touch on legal and ethical issues that directly affect the way we do our work as usability professionals.

In upcoming columns I'll tackle issues such as the use of non-disclosure forms, video releases, IRS regulations on the payment of test subjects, copyright issues regarding user-interface designs, and the like. Although this column will not provide legal advice per se, I hope it will heighten your awareness of the legal implications of much of the work we do. Occasionally, I intend to use this space as a forum for attorneys to discuss relevant legal issues affecting usability professionals.

In this inaugural column, I want to urge that the UPA develop and promulgate a code of ethics for its membership. Because we deal directly with other human beings in our testing, it is critical that our members agree on what constitutes ethical and unethical behavior, and use such code as the basis for their usability testing activities. There is substantial precedent for taking this step: All human or animal-based scientific professions have adopted remarkably similar and specific guidelines for the treatment of human beings and animals in test situations. Further, we have already taken steps to formalize our informal association through by-laws and dues. An ethics code moves us one step closer toward becoming a full-fledged discipline.

These are the basic tenets I urge the UPA to adopt with regard to usability testing. I welcome members' comments and suggestions for additional items:

All usability professionals are responsible for the ethical treatment of all participants in a usability test.

- Usability tests must be designed and executed so as to prevent physical or psychological harm to the participant. Following a test, the participant should rightly expect to be in a physical and mental condition comparable to that existing prior to the test.

- Prior to a test, participants must be apprised of their rights as test subjects: the right to anonymity, the right to stop the test at any time and for any reason without forfeiting the pay or compensation agreed upon, and the right to ask questions about the test at the end of the test.

- Participants must formally agree (in writing) to being audio- or videotaped or photographed, and must be apprised of the uses for which those recordings or photographs may be used.

- When deception must be a critical element of a test, e.g., the participant is told that the usability professional is from a fictitious organization to prevent biasing the test results, the participant must be told of the deception during a post-test debriefing.

- All participants must be provided a post-test debriefing to enable them to ask questions or report any physical effects resulting from their participation in the test.

- Participants should ordinarily be compensated for their participation except in instances where their employer forbids such compensation, or where compensation could bias the data collected.

Methods In Usability

Usability Throughout the Product Development Cycle

Elissa Darnell, WebTV, and Shannon Halgren, Sage Research

Introduction

Traditional usability testing, in which representative users are observed while performing representative tasks has become quite common. In fact, it is fairly commonplace for a computer company to have a usability lab that is designed to optimize observation and recording of usability testing. However, the current challenge which appears to face today's usability practitioner is moving beyond the use of a single user research method—the conventional high-fidelity usability test which is typically conducted too late in the development cycle to have significant impact on design. In our tutorial given at the 1997 UPA conference, we discussed a variety of usability research methods that can be used throughout the product development process. In this short article we cannot discuss all of the techniques that we were able to cover in our one-day tutorial.

Instead, we would like to introduce a few of the less common techniques that can be used in the early pre-design phase and in the late future-planning phase of product development. These techniques can increase the usability practitioner's involvement during product development and therefore increase the overall usability of the product.

The design process

While each computer company may use its own jargon for the steps in their software development process, the overall development process is highly similar across computer companies. A typical product development cycle is shown in figure 1.

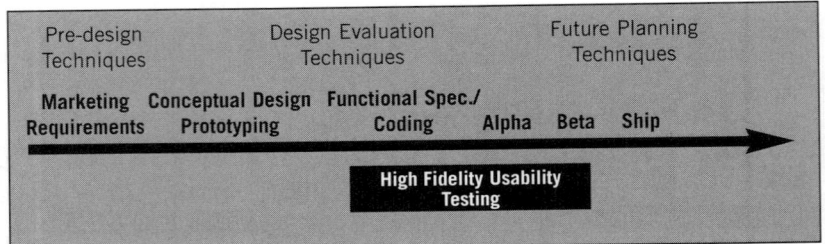

Figure 1: The typical development cycle and where usability testing usually occurs.

A product typically goes through three broad phases during its development. In the pre-design phase, the marketers are active setting feature requirements for the product. During the design phase, drawings or prototypes are developed demonstrating the look and behavior of the product. The product coding then continues through the stages of alpha and beta, and is eventually shipped. Prior to shipping and in between product cycles, the product team begins planning for the next release.

As is apparent from figure 1, the time in which high-fidelity usability testing is often performed is relatively late in the development cycle. When conducted this late in the development process, the study is forced into focusing on small design refinements rather than the product's fundamental conceptual design.

To increase the effect user research can have on a product requires expanding the scope of user research during the development cycle. Usability practitioners can contribute to the fundamental design of the product and even to marketing requirements by conducting user research in the pre-design phase. They can also help plan for the next release of the product by using user research techniques on the beta and shipped version of the product in the future planning phase of the development cycle. Figure 2 shows a variety of methods that can be used throughout the development cycle.

Since many usability practitioners are aware of, and comfortable using, design evaluation techniques such as usability testing and usability walk-throughs, we will focus more on the pre-design and future planning techniques that are less commonly used, and broaden the scope of involvement in the development cycle.

Pre-design techniques

Before work begins on the initial design of the software, there are several

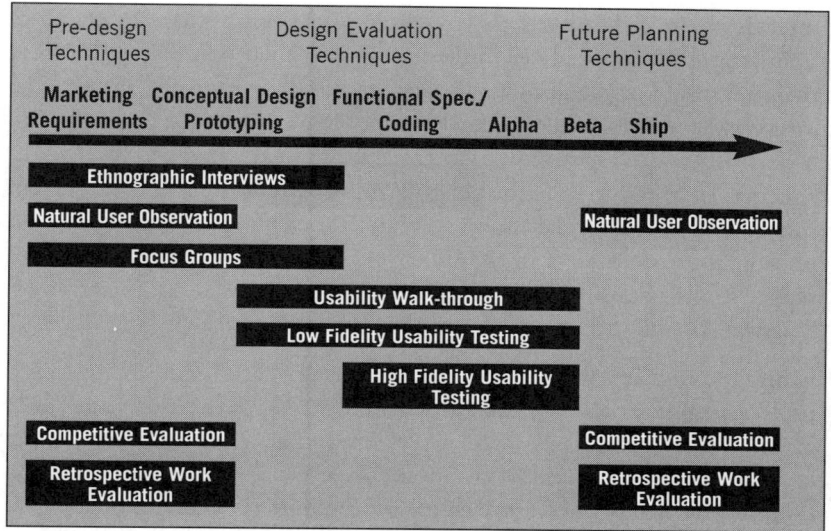

Figure 2: The typical development cycle and the variety of usability research methods that can be applied throughout the development cycle.

techniques that the usability practitioner can perform which can contribute to the initial designs. These techniques are ethnographic interviews, natural use observation, and focus groups. All three techniques focus on understanding the user and their tasks, how users currently perform their work, and what they currently find frustrating or limiting.

Ethnographic interview

The ethnographic interview is a research method used in anthropology. Ethnography is an attempt to understand and describe a specific culture or group. Anthropologists achieve this understanding by conducting well-crafted, in-depth interviews and hours of observation. The usability practitioner can apply a modified version of this research technique to gain an understanding of the user's culture/environment, work processes, work flow, tasks, and responsibilities. The data collected from ethnographic interviews can be used to form both a workflow model and a physical model of the user's environment. This information can contribute to the marketing requirements and design of the product.

There are many ways to conduct an ethnographic interview. One example of an interview might begin by asking the interviewee to list his or her tasks or job responsibilities. If interview time is limited, the interviewee may be asked to rate tasks or responsibilities in terms of frequency and importance. These ratings can then be used to structure the interview. For each important and frequently performed activity, the interviewee may be asked to describe and demonstrate the processes and tools (software, copy machine, forms, etc.) used at a detailed level. They also may be asked to describe delights and frustrations with their current processes and tools. The usability practitioner may also collect or take photographs of artifacts in the interviewee's environment. Artifacts can range from examples of the output of users' tasks to forms used, diagrams drawn, photos of planning boards, etc.

Ethnographic interviews can provide rich data and an in-depth understanding of the user and his/her work. However, this abundance of data takes a great deal of time to review and analyze. It can sometimes be difficult to find commonalities across users. Keep in mind, a good sampling of users is necessary to make sure data is generalizable.

Natural use observation

Natural use observation is a technique that nicely complements ethnographic interviews and is therefore often done in conjunction with it. This technique involves observing users doing real work in their work environment. While the ethnographic interview focuses more on actively interviewing the user, natural use observation is one of passive observation. A usability practitioner can combine the two techniques by alternating between periods of passive observation and active question asking. It is important, however, to have at least some times of passive, fly-on-the-wall observation to get a realistic look at the user's processes.

Observation is a way to learn about users' needs and workflow. Often it may also result in the discovering of a user's unanticipated usage of software. Two ways to perform natural use observation are to 1) directly observe the user as he/she works, and write down observations, and 2) indirectly observe the user by videotaping. Passive natural use observation is extremely time-efficient because it does not require the user to take time out of their work day to participate and it requires little researcher preparation. It is also ecologically valid and provides rich data. However, it suffers from the same trade-offs as ethnographic interviews in that it may be difficult to discover commonalities across users, and it requires extensive time to analyze the written notes or videotapes.

Focus groups

A focus group is another effective pre-design technique. It involves having a group of users (approximately six to nine) discuss new concepts and identify issues over several hours. Session activities can range from group discussion to drawing interfaces. Focus groups can elicit users' needs and attitudes. The focus group is a research method common to usability practitioners as well as market researchers. However, we would like to describe two variations on the traditional focus group technique that offer some additional advantages and flexibility.

During an electronic focus group, users participate in the focus group session via an electronic "chat room" rather than in person. The focus group is still run by a facilitator who moderates the discussion and keeps the group on track, but does so remotely. This variation allows users to participate from all over the world without the expense of travel. Another advantage of this type of focus group is that participants type in their comments rather than verbalize them, allowing a record of the session to be automatically generated from the participants' comments. This facilitates quicker turnaround time for the analysis. However, the electronic style of interaction lacks the conversation cues present in face-to-face discussion, which can be informative. The absence of these social cues results in multiple people entering their comments simultaneously, making it difficult to follow a thread of discussion. It's almost as if all the participants are talking at once! There may also be interference from technical problems such as the chat room interface may impose a limit on the number of lines a user can enter in a single comment. During our personal experience conducting an electronic focus group, the group facilitator's modem connection was interrupted and the facilitator was temporarily dropped from the session.

A second variation on the traditional focus group is the use of a computer-enhanced meeting room. This is an enhancement of the facility and tools used to conduct a traditional focus group. The group of users meets in a room that has a series of networked computers. Each session participant enters his/her responses and comments via a computer rather than verbally. The agenda for the session, questions and topics of discussion, are communicated to the users through groupware software. This software is designed for collaborative work and group decision making. During a session, users may be shown several prototypes demonstrating alternative designs and asked to rate or rank order their preferred design. They may also be asked questions about the type of tasks they perform, and features they need. An example of a facility like this can be seen at the IBM Santa Teresa Laboratory in San Jose, California. The advantages of using a computer-enhanced meeting room like this one is that it allows for independent and anonymous user input. In a traditional focus group session, some users may be more verbal than others and a dominant personality may end up influencing others in the group. Another advantage is it also results in an automatic record of the session which reduces transcription and data-collection time. In fact, the groupware software can be used to analyze group votes during the session and open the results up for discussion with the group. However, this technique is not without trade-offs. This interaction style may discourage or inhibit spontaneous group interaction and discussion. Since the session moderator does not see the user's response immediately, he or she cannot probe for more information or encourage further elaboration.

Post-design/future planning techniques

Future planning techniques are research methods conducted on beta software or completed commercially available software. Just as the automobile industry releases new models of their cars every year, the computer industry continually refines, upgrades, and enhances its product releases. Given this standard of version releases, it is important to use the currently available products (either your own or a competitor's) as the starting point for the next release of your product.

Retrospective work evaluation

Retrospective work evaluation is a technique performed with users who have been using the product in question for some period of time. The evaluation session is conducted where the user uses the product. The session begins by asking the user to share work previously created with the product. They are also asked to explain how the work was done. Sometimes the researcher will ask the user to redo the work so that the researcher can observe exactly how the work was done. This will also serve to jog the user's memory about difficulties that they may have encountered when initially creating the work.

This technique can provide a realistic view of how the product is being used. Because all participants have been using the product for some time, it can also uncover long-term usability problems rather than strictly initial-use problems. This is something traditional usability testing has difficulty eliciting. Retrospective work evaluation can also shed light on user frustrations and product limitations, as well as getting user comments based on real work experience.

When using this technique, it's often helpful to ask the user to prepare for the session by finding examples of work they have created with the product. This avoids the wasted time watching users search for past work. The researcher should have the user walk through how the work was created and ask to see the user re-create areas of interest. Areas of interest may include work that was particularly difficult to create or work that required the use of specific features of interest. The researcher should ask the user about delights and frustrations about the process and collect work samples. This technique is advantageous in that realistic data regarding product usage and context can be collected. Rich data can be collected with little time wasted gathering irrelevant data. However, some limitations of this technique are that information on first-time use is not gathered and user's memories about the work creation process may not always be accurate. Additionally, the work seen is limited to what users select which may be only the projects they're most proud of and not the most typical work examples.

Competitive evaluation

Competitive evaluation is a technique for comparing the usability of your product with its main competitors, what it will replace in the marketplace, or another idea for a new interface. The evaluation can be conducted by usability practitioners or by users. Through a competitive evaluation, you can learn the strengths and weaknesses of each product, which product design has performance advantages, and which features users prefer.

There are many ways to conduct a competitive usability evaluation. One approach is to select features that both products claim to support, and create realistic tasks based on them. Then have the user complete all the tasks on first one product and then the other product. It is important to counterbalance the product order and to randomly assign users to each order. This helps to minimize any learning or carry-over effects from one product to the other. Objective measures such as time, accuracy, task completion, help referrals, and ease-of-use ratings can be collected.

Subjective questions regarding user preference can be asked after each task or session is completed. This technique provides good, direct comparison data and can be used to learn from other products strong and weak points.

Conclusion

In this article we have discussed some variations on traditional usability techniques that we have used in our work, and demonstrated how usability methods can and should be integrated into the product development cycle. We hope that, over time, usability methods early and late in the development cycle will become more commonly used and will result in greater improvements to the usability of software products.

Some helpful references about the techniques discussed

Bauersfeld, K. & Halgren, S.L. (1996). "You've got three days!" Case Studies in Field Techniques for the Time Challenged." *Field Methods Casebook for Software Design,* D. Wixon & J. Ramey (eds.). New York: John Wiley & Sons.

Fetterman, J. M. (1989). *Ethnography Step by Step.* Newbury Park, CA: Sage Publications.

Diaper, D. (1989). "Task observation for human-computer interaction," *Task Analysis for Human-Computer Interaction.* Diaper, D. (ed.). Chichester, UK: Ellis Horwood, pp. 210–237.

Caplan, S. (1990). "Using focus groups methodology for ergonomic design." *Ergonomics.* pp. 5, 33, 527–533.

Goldman, A. E., and McKonald, S.S. (1987). *The Group Depth Interview: Principles and Practice.* Englewood Cliffs, NJ: Prentice Hall.

Greenbaum, T. L. (1988). *The Practical Handbook and Guide to Focus Group Research.* Lexington, MA: D.C. Heath & Company.

Brooks, P. (1994). "Adding value to usability testing." *Usability Inspection Methods.* Nielson, J., and Mack, R.L. (eds.). New York, NY: John Wiley & Sons.

Brevity Versus Usability: On the Need for Articulate Messages in User Interfaces

Howard Tamler,
HT Consulting

I begin with the observation that most of the content of "graphical" user interfaces is conveyed by text, not graphics, so that the effective use of language is basic to the way that user interfaces communicate with users. In particular, it seems to be a pervasive tradition in software design that the text in user interfaces tends to be as brief as possible. One likely historical reason for UIs being so terse is that in previous decades when computers had much less capability, the RAM needed for displaying text was much scarcer than it is now, and low-resolution monitors with small screens had a serious shortage of screen real estate for displaying text or anything else. Another likely reason is the traditional promotion of brevity in the usability guidelines literature.

For example, Nielsen's (1993, p. 20) first heuristic guideline, "Simple and natural dialog," states that, "Dialogues should not contain information that is irrelevant or rarely needed. Every extra unit of information in a dialogue competes with the relevant units of information and diminishes their relative visibility." Of course, the key words here are "needed" and "relevant." Likewise, a cursory search of Mayhew (1992) turns up at least four prescriptions for brevity: "Menu choice labels should be brief" (p. 151); "Captions should be brief" (p. 193); "Prompts should be brief" (p. 204); and "Provide brief prompts and instructions" (p. 225). However, the examples for each of these prescriptions differ only in the presence of extraneous, not essential, words.

In line with this, Microsoft's (1995) guidelines for menu item labels say to "keep the wording brief and succinct," because verbose labels make it harder to scan; similar statements appear in several places in these guidelines. In all fairness, this guidelines document is sensitive to the trade-offs; that is, it generally pairs these directives with prescriptions for avoiding truncation or for making the message clear, descriptive, and complete. To compensate, it also suggests using "tooltips," as well as small info boxes invoked by the "What's this?" command on right-click contextual pop-up menus.

On the other hand, there are equally good arguments that brevity guidelines have been overly successful, and that some corrective swing of the pendulum is needed in order to restore an optimal balance on this issue. For example, Reeves and Nass (1996) point out that most menu systems describe each option with only one or two words at most, regardless of the complexity of the function; consequently, users are frustrated by not getting the whole story. They argue that the "use of plain English (full sentences or at least multiple words in logical phrases) would make an enormous difference in understanding and satisfaction" (p. 30). Elsewhere, describing Microsoft Bob, the first commercial attempt to implement a "SUI" (social user interface), Nass has argued that users appreciate complete sentences and prefer them to sentence fragments, since this is how people talk in real life. Consequently, all the Microsoft Bob characters speak in full sentences, which is more polite and sociable than the one-word utterances so common to UIs.

As another example, *User Interface Engineering* (1996) has reported evidence of the need for increased verbosity. Specifically, the authors compared the "tooltips" of Microsoft® Access 2.0 with the analogous "hover bubbles" of Lotus® Approach 3.0. The Microsoft tooltips are very brief—for example, the icon which is redundant with the "Form" command on the "View" menu has a tooltip which says "Form View." In Lotus Approach contrast, however, the hover bubble for one of the Lotus Approach icons which is redundant with the "Browse" command says, "Go to Browse to review or modify data." In other words, the Microsoft tooltip merely names the command, while the Lotus bubble not only names the command, but also provides the reason for using it. Clearly, in this example, the added verbosity seems to provide a level of description which goes significantly beyond what's offered by the Microsoft tooltip. Most important, the additional words seem to help— users found Approach easier than Access in terms of finding functions quickly and understanding the icons and menus.

Likewise, *User Interface Engineering's* (1997) comparative study of various Web sites found that longer links work better and that better Web sites (e.g., the Edmund's automobile shopping site) have longer, more articulate links. As the authors put it, "The Edmund's site finished first in our study. . . . We believe this is due in part to its long, descriptive links," and "The Hewlett-Packard site finished third in our study. Part of its success may lie in how different its links are from each other." Their common-sense explanation is that something about the additional words seems to have helped users pick the links that led to the desired information. More specifically, the effectiveness of a link is a function of how easily the user can predict where the link will lead and how easy it is to discriminate a given link from other associated links, both of these tasks being enhanced by the addition of information. In addition, I think it's important to point out that articulate messages may be more important on the Web, where textual content is the main feature, as opposed to productivity software in which functionality is the main feature.

In my own usability work, I have come across many instances of confusing UI text (menu items, command buttons, field labels, and especially error messages) whose meaning could have easily been made clear by adding a few words or otherwise spelling out the utterance more completely. In addition, while in some cases ambiguities can be resolved by the context, this context may not be readily understood by a novice user. It is hard to come up with good examples because any given illustration seems rather trivial by itself. However, while none of the following four examples is earthshaking, the cumulative effect of many such instances could have a significant impact on the clarity and credibility of an entire system.

1. In a usability test of an optical spectrum measuring instrument, several users (all of whom were instrument-using engineers) misunderstood the "Other Traces OFF" command button. They didn't know if it meant that the traces are already off, or that this button would turn them off; several thought that it means that traces are now off, and that if they click it, it toggles to "Other Traces On." I recommended that this ambiguity be resolved by relabeling it "Turn off other traces," which required adding only one key word.

2. In another product, the message "OK to delete" is displayed when a selection is made, meaning that deletion can be executed by pressing the "OK" key. Given the context, this could be misinterpreted as "it's OK to delete the thing that you selected." Consequently, I recommended changing the message to "Press OK to delete," an addition of only one word.

3. In a usability test of a customer profile application, one task required users to search for customers within a specified geographical area. When users were looking for all customers in Paris, France, some were repeatedly confused by an error message saying, "Search criterion missing." On the face of it, this message was wrong, since "Paris" was a perfectly valid search criterion. What these users didn't know yet was that the system would not search on a city alone, but required that users also enter a customer name or alias, as well as a city. Hence, what the message really meant was that at least one, but not all, of the required search criteria was missing. I recommended that the message articulate the problem with something specific such as, "You must enter a customer name or alias in addition to a city."

4. I evaluated a financial application where the following text appeared in a dialog related to income tax:
   This tax tracking type is usually associated with deductions. Report elective deferrals up to the annual limit of $10,000. [Product] reduces amounts on Form W-2, Box 1 and Form 941, Line 2. Checks "Pension Plan" and "Deferred Comp" checkboxes on Form W-2, Box 15. [Product] reports on Form W-2, Box 13, Code D."

While this explanation was potentially helpful, the "sentences" are somewhat clipped and compressed, and as a result, unclear in the following ways:

• Who is supposed to "report elective deferrals . . ."—[Product] or the user? Who "checks 'Pension Plan' and 'Deferred Comp' checkboxes . . ."—[Product] or the user? The lack of a subject in these sentences makes them ambiguous as to whether they are imperative or descriptive, and as to who or what is doing the reporting and checking.

• The last sentence is ambiguous, in that it could mean that [Product] enters some value on Form W-2, Box 13, Code D or that the [Product] reports include the value which can be found on Form W-2, Box 13, Code D.

a. reduce the amounts on Form W-2, Box 1, and on Form 941, Line 2, by the amount of the deduction.

b. check the "Pension Plan" and the "Deferred Comp" checkboxes on Form W-2, Box 15.

c. enter Code D on Form W-2, Box 13.

So, what makes longer UI labels, links, messages, and text panels easier to understand? Foremost in any explanation is the psycholinguistic fact that many—probably most—utterances are ambiguous, but we routinely interpret them only one way because the context, especially the unique combination of the words in the utterance, makes a particular interpretation more plausible. More important for UI design, which traffics in short phrases and single words, Glucksberg and Danks (1975) argue that single words such as "pen" are typical in having a multiplicity of meanings, since linguists agree that ambiguity in single words is the rule rather than the exception. While some think that at least 50 percent of the English lexicon is ambiguous, others think that practically every word is ambiguous in principle, especially if we include metaphorical usage.

The other part of this equation is that ambiguity is inversely proportional to the length of the utterance, all things being equal—that is, each additional word in an utterance, analogous to a criterion in a Boolean search, further constrains the space of possible meanings. Hence, a word is easier to interpret in a sentence or phrase than in isolation, and longer phrases tend to have a more specific meaning than shorter ones. For example, the famously misinterpreted phrase "Press any key" would be clearer if it said "Press the Enter key" or "Press any one of the letter keys." Likewise, the label "display mode" can mean "the current mode of display" or "display the current mode"; however, it takes only one additional word to disambiguate: either "display the mode" or "mode of display." As another example, the word "take" has so many meanings, both verb and noun, that it takes up three-quarters of a page in Webster's dictionary; while "take my wife—please" plays on this ambiguity, but "please take out the garbage" is unambiguous.

Coming at this from a similar angle (namely, Claude Shannon's

Information Theory, Tognazzini [1992]) makes an analogous claim about the clarity of meaning as a positive function of the amount of textual information. Not only do the rules of English grammar (including word probability, spelling, etc.) describe a language in which over 75 percent of the information is redundant, but this is more or less true for every language that's been studied. Why should this be the case? Because human communication always has a rich potential for error or misinterpretation arising from a variety of sources: the message content, the transmission medium, the signal/noise ratio, the context, the receiver's assumptions, etc. Hence, it's only thanks to the significant redundancy of language that receivers are able to get the message right. (A familiar application illustrating this principle is the standard postal address, where the ZIP code is entirely redundant, but nevertheless useful.) The upshot is that some level of redundancy is good for user interfaces in particular, as well as for human communication in general, and additional information in a UI should not be ruled out merely because it might be redundant.

In conclusion, I'm not advocating any universal principle that verbose is better; rather, the optimal message length varies with the particular situation and is typically somewhere in the middle rather than at either extreme.

I am merely claiming that, contrary to many UIs and the pitch of some UI design guidelines, the shortest message isn't necessarily the most usable; and that in many cases, usability can be enhanced by lengthier, more explanatory textual messages. Moreover, while brief messages are appropriate for experienced users of a system, relatively detailed messages with more redundancy may be better for novice or infrequent users. Since these judgment calls can be difficult, it may sometimes be useful to have a technical communication specialist on the usability team. And with dynamic GUI devices such as tooltips, balloons, and pop-up panels, it need not cost us anything in terms of valuable screen real estate to provide more clear, complete, and articulate messages.

References

Glucksberg, S. & Danks, J.H. *Experimental Psycholinguistics.* Hillsdale, NJ: Lawrence Erlbaum Associates, Inc., 1975.

Mayhew, D.J. *Principles and Guidelines in Software User Interface Design.* Englewood Cliffs, NJ: PTR Prentice Hall, Inc., 1992.

Microsoft. *The Windows Interface Guidelines—A Guide for Designing Software.* Microsoft Corporation, 1995.

Nielsen, J. *Usability Engineering.* Cambridge, MA: Academic Press, Inc., 1993.

Reeves, B. & Nass, C. *The Media Equation.* Stanford, CA: CSLI Publications, 1996.

Tognazzini, B. *Tog on Interface.* Reading, MA: Addison-Wesley Publishing Company, Inc., 1992.

Spool, J.M., Scanlon, T., Schroeder, W., Snyder, C., & DeAngelo, T. *Web Site Usability: A Designer's Guide.* North Andover, MA: User Interface Engineering, 1997.

User Interface Engineering. "Effective Tool Tips," *Eye for Design,* 3 (1), 1996: pp. 1–7.

An Application of the Principles of Minimalism to the Design of Human-Computer Interfaces

JoAnn T. Hackos,
Comtech Services, Inc.

This paper was originally presented as a keynote speech at the German HCI conference, Software-Ergonomie '99.

Minimalism in information design, specifically as applied to user tutorials and manuals, was introduced in the early 1980s through the work of Dr. John M. Carroll, then a cognitive psychologist at the IBM Watson Research Center. Since that time, theorists and practitioners have further elucidated the principles of minimalism and have attempted to apply it to a variety of situations in which people attempt to learn how to use a software application. Most recently, a new exposition of minimalist principles and practices was published by MIT Press. This work, *Minimalism Beyond the Nurnberg Funnel*, represents the work of leading theorists and practitioners in the field.

I have long been in the habit of describing the user interface as an element of information design and thus, amenable to the basic design tenets underlying information design. And, I have frequently characterized the user interface as an expression of minimalist design because the interface combines the traditional design elements of text, graphics, and layout in a two-dimensional space, with the occasional addition of movement, in as compact a form as is practical. More recently, I have been intrigued by the possibility of applying minimalist principles to interface design in a more systematic manner than has been heretofore discussed by its proponents.

In this presentation, I will introduce the four basic principles of minimalism, as well as some of the research done to support the principles, and explain how they might contribute to our understanding of interface design. To the extent that time and space permit, I hope to present some examples of how a minimalist interface might be similar but also differ from more traditional interfaces.

Background to the minimalist debate

John Carroll's most developed presentation of the minimalist concept appeared in 1990 in *The Nurnberg Funnel* (MIT Press). In this work, he fully described what he and his team at IBM had learned from observing people trying to learn to use IBM's DisplayWriter word processor and how to learn the computer language, SmallTalk. Learners of DisplayWriter and SmallTalk were most successful when they were provided with brief instructional material that encouraged them to act rather than read. They learned best when the tutorials emphasized the goals they really wanted to achieve, rather than tasks defined by the computer software and the interface.

In one case, for example, the researchers created a cue card that explained to the secretaries how to "type something" rather than how to "create a document." "Creating a document" was the heading used on one of the on-screen menus (before Windows) and in the actual DisplayWriter documentation. They had discovered that the secretaries did not want to create documents and were unable to relate their goal, which was to type something, to the name of the task in the interface. This mismatch between goal and execution of the task was a significant impediment to their learning.

In the design of the SmallTalk tutorial, the researchers discovered that programmers were best able to learn the computer language when they were engaged in the actual tasks of writing and debugging code rather than reading about the concepts underlying the language. Users who are encouraged to act, rather than read about acting, are more successful in their learning. Users who are encouraged to perform tasks that are directly related to their goals in using the software are more able to formulate realistic plans, the set of steps they envision will take them through a series of actions to their goals.

Minimalist principles

In 1996, John Carroll and Hans van der Meij summarized the principles of minimalist design of documentation and training in an article in

Technical Communication, the journal of the Society for Technical Communication. This article, reprinted in *Minimalism Beyond the Nurnberg Funnel*, most clearly states the four minimalist principles and how they are represented in the design of documentation and training.

The four basic principles of minimalism are:

Principle 1: Choose an action-oriented approach

Principle 2: Anchor the tool in the task domain

Principle 3: Support error recognition and recovery

Principle 4: Support reading to do, study, and locate.

My purpose here is to examine to what extent these principles, used to define goals for the development of documentation and training, apply to the design of the human-computer interface.

Principle 1: Choose an action-oriented approach

The first principle of minimalism points to the basic concept that underlies interface design. An interactive interface implies action; the user is provided with opportunities for action. In a Windows environment, the user selects commands from a task bar, makes selections in dialog boxes, types information into data-entry fields, and so on. Interfaces ordinarily provide many opportunities for action, even if that action is simply looking at an on-screen report. The intent of basic interface design is to allow the user to do something.

However, there is more to the first principle than simply action. An action-oriented approach implies that the user is able to accomplish something. The user is most satisfied, it appears, when they have an immediate opportunity to act and when they believe that the actions they take will lead them toward their goals. Simply clicking on buttons or typing something into a field does not imply that the actions are anything more than random attempts to make progress. In many flawed interfaces, the actions that the user takes often have little purpose. They are simply attempts to see what, if anything, will result. We observe users trying to guess, unsuccessfully in many cases, which actions will lead to the results they want. The interfaces provide them with little or no guidance about where to start.

I recently reviewed an interface design in which the user is presented with a blank screen and a task bar when they enter the program. Their first cognitive task is to examine the task bar and attempt to guess what item they need to select to accomplish something. The user is immediately confused, unable to take the immediate and purposeful actions that they want to take, because the interface fails to lead them in a clearly defined direction. A more successful design, following minimalist principles, would present an immediate opportunity for action in the context of the user's goals. For example, if the user needs to select a patient in order to view that patient's record, the first action available immediately in the interface should be patient selection. Providing an immediate opportunity to act means making the desired activities immediately apparent and reducing the number of choices that the user must make before taking action.

In addition to an immediate opportunity to act, the first principle of minimalism also points to the need for the user to explore as a way of learning. Most graphic user interfaces provide more than enough opportunities for exploration, at least in principle. The standard Windows approach is to make many actions available through task bars, pull-down menus, and dialog boxes. The users can browse through the menus, select and review dialog boxes, and make different choices to see how those choices might affect what is happening on the screen. The problem with this approach is that the exploration is undirected. We have found that when the user explores at random in an interface, they frequently become increasingly confused, unable to differentiate among explorations that lead to their goals and explorations that lead away from their goals. An interface that guides exploration in productive ways leads the user by using information on the screen that makes clear the direction to take.

Finally, the first principle implies that we must, in creating successful designs, respect the integrity of the user's actions. When we present message boxes that are unrelated to the user's actions, we violate this aspect of the first principle. When we present error messages that fail to inform the user how to correct the problem, we also fail to respect what the user is attempting to do. Tips that appear on the screen, unasked and unwanted, distract the user from their tasks. Tips that attempt to track user actions have the promise of maintaining the user's original intent, but, thus far, seem mostly clumsy and overbearing.

One of the researchers at Xerox PARC remarked to me a few years ago that they were having a difficult time making sense of the user's seemingly random keystrokes to approach a task. They could not anticipate the user's intent simply from monitoring keystrokes. The researcher suggested that a much more sophisticated approach to interpreting the user's logic was needed before such tracking might be successful.

For an interface to comply with the first minimalist principle—to take an action-oriented approach—means the following:

- Make critical actions immediately apparent when the user enter the interface, especially as the user enter the interface for the first time.

- Provide opportunities for exploration through the interface, but guide those explorations using appropriate text and graphics so that the connection between actions and goals are easily apparent.

- Respect the integrity of the user's actions by eliminating annoying and often gratuitous invasions in the middle of tasks the user is attempting to pursue.

Principle 2: Anchor the tool in the task domain

The second principle is most significant for the design of user-centered systems. Too often, interfaces reflect the underlying database structure rather than the user's goals. To anchor the tool in the task domain means to design the interface from the user's perspective rather than designing an interface that is functionally correct but disembodied. We have all seen interfaces that fail to resonate with the user's goals. These interfaces have functional names that make no sense to the user. They require sequences of actions that are hidden from view, requiring the user to figure out which task structure will lead them to their goals.

For example, we recently reviewed a Web site designed by a major American university. We took the point of view of a user who wanted to find out if the university offered courses toward a degree in technical communication. We believed this goal to be typical of many individuals coming to the Web site from outside the university. Unfortunately, the Web site was not designed from the user's perspective, but rather used the organizational structure of the departments and schools in the university to organize the information. Unless the user already knew which school and department offered the courses he/she wanted to fit, they could not get the information needed. In fact, he/she needed to know the solution to their problem in order to find the solution—a rather circular approach unlikely to produce success.

In order to anchor the tool in the task domain, we must first be well informed about the goals that the user wants to achieve. Do they want to use the task bar or send a fax to a colleague? Do they want to complete the dialog box or find out how many days of vacation they have left? In many interfaces, the user's real goals are obscured, if they were ever understood in the first place. We need to use all of the user-centered design tools to ensure that we truly understand the world from the user's perspective if we hope to build successful interfaces.

The second principle of minimalism exhorts us to build on the user's prior skills, knowledge, and experience in the design of product. As a consequence, we must know the user's goals and relate them to tasks already known in the user's domain to achieve those goals.

For example, the software program "Quicken" uses a metaphorical construct on screen of a check register, similar in appearance to the check register provided by banks with packets of checks. By using this visual metaphor of the task, the designers are able to build upon the user's knowledge of performing the task with the physical artifact. The affordances of the artifact are transferred to the interface. The user is able to bring their prior experience of recording checks and balancing their checkbooks to performing the task in the new environment.

Metaphoric constructs alone, however, are not sufficient to reinforce the user's prior skills, knowledge, and experience. In many more complex system designs, the overall relationship among tasks must reflect the user's experience. For example, consider a system to support the task of call tracking, in which the overall flow of information from the primary screen, through secondary screens, and into dialog boxes, is made to resemble the task sequence performed by the user in the physical environment. If we are able to take advantage of known sequences, we spare the user the need to unlearn their previous behaviors and learn new ones. Their cognitive tasks are simplified, the learning curve is reduced, and they are able to accomplish their goals.

In building an interface that respects the second principle of minimalism, we must place information in the interface that supports task performance and enables the user to link the tasks to their goals. To do so often requires more than cryptic terminology. For some reason, we appear to have chosen to preserve the limited language present in the original teletype interfaces. Menu systems often consist of single words when phrases might be more meaningful. Dialog boxes reduce text to field labels and one-word descriptions of radio buttons or check boxes, often making it difficult for the users to know what actions to take.

Given the space we have to work with and the legibility of a graphic user interface, we appear to be constrained by unnecessary brevity, afraid to use words to help the users know what is going on. If we are to support the user's task performance, we should consider adding meaningful language to the screen, using text in phrases, sentences, or even paragraphs. In the past few years, interface designers have introduced the concept of the wizard or coach to assist users in performing certain tasks. Sometimes those tasks have been designed so that novice users are better supported, in effect, supplying an alternative to the more cryptic standard interface. In other cases, we have designed wizards to become the primary interface for complex tasks that users have difficulty performing successful with the cryptic assistance given by the "ordinary" interface. Performance support of these types have proved inordinately successful in improving task performance and enabling users to learn how best to perform tasks and achieve their goals.

It is interesting to note that, in most instances, the performance assistants (wizards and coaches) carry more text explanations than do the "ordinary" interfaces. If we have learned that more text can result in better performance, in the context of a well-designed task environment, why not adopt such devices most of the time instead of making them special cases?

In general, on-screen text should provide the user with clues about the task structure in the software. Terminology used on the screen should reflect the user's goals and lead them through the structure of the tasks. The text on-screen, as well as the layout of the information (reflecting reading order or other structures), should help the user make the connection between what they want to do and how they accomplish the tasks within the software application.

For an interface to comply with the second minimalist principle—to anchor the tool in the task domain—means the following:

- Ensure that the user's goals are well understood by the interface designers so that the interface makes a clear connection between the user's goals and the tasks required to achieve the goals.

- Build on the user's prior skills, knowledge, and experience by creating a metaphoric structure for the interface that enables the user to connect known abilities to new requirements for interaction.

- Place information into the interface that supports task performance. Consider using more rather than less text in the interface, along with layout, color, and white space to assist the user in learning how to perform the tasks within the tool.

Principle 3: Support error recognition and recovery

The third principle of minimalism emphasizes the user's need to detect the problems when they occur, diagnose what has happened, and find effective ways to overcome those problems. At the heart of this principle is the need to prevent mistakes before they occur. Often, preventing mistakes requires careful use of default values for fields and choice buttons and error trapping. For example, I was reviewing an interface that required the user to type the hyphens in their U.S. Social Security number (the standard format is xxx-xx-xxxx). If the user failed to include the hyphen, a message appeared stating that the hyphens must be typed. In addition, all the numbers the user had already typed were deleted from the field and replaced with a blank space. Such an approach is insulting to the user and reflects sloppy programming by the developer. First, there is no need to require the hyphens at all. If they need to be added, they can be programmed to appear automatically as the user types. Second, if a particular sequence of characters is required in a field for some reason, and the user makes an obvious mistake, the cursor should be returned to the position to correct the mistake following the appearance of the error message.

In many cases, as we have all seen, error messages are both insulting and belligerent toward the user. The developers, in effect, accuse the user of being careless and stupid. The developers, in many instances, have chosen to blame the user rather than do the additional programming to prevent the error in the first place.

I am frequently annoyed when I enter a dialog box and have to establish the keyboard focus by clicking the cursor in the first field. Some additional programming is required to place the cursor in the first field, but someone considered that additional programming a waste of time. A decision to waste the time of hundreds, if not thousands, of users to save minutes of programming time should not occur.

In areas where errors are likely to occur or when error correction proves difficult, we need to design-in cautions and warnings to alert the user about the potential problems and how to recover. If we provide error messages in such cases, the error messages must be carefully worded to assist the user in detecting the error, diagnosing what happened, and taking corrective action. Too often, error messages simply state the nature of the error, often in rather cryptic language, rather than providing assistance for correcting the error.

One way of assisting the user to avoid errors is to provide many default values and explanations within the interface about what might be expected if new values are selected. Warning messages that point to the possibility of future problems are also useful additions, especially when they provide sufficient information to allow the user to make choices about their course of action.

For an interface to comply with the third minimalist principle—to support error recognition and recovery—means the following:

- Assist the user in preventing the error in the first place by using error-trapping techniques as thoroughly as possible.

- Assist the user in making the right decisions by providing default values and complete information about the choices available in the interface itself.

- Create informative, helpful, and courteous error messages that enable the user to detect, diagnose, and correct problems as soon as they occur.

Principle 4: Support reading to do, study, and locate

The fourth principle of minimalism—to support reading to do, study, and locate—applies primarily to the help system and documentation that you design to support the interface. We know that most of the time, the user wants to read to do (Redish, 1988), seeking information that helps them complete the tasks in the software application and reach their goals. Help systems and documentation that emphasize reading to do, providing task-oriented instructional text, are most likely to assist the user in navigating through the application interface.

Users who focus on reading to do profit from context-sensitive help and embedded help systems. Context-sensitive help is designed to anticipate the information the user may need at a particular point in the interface. If the help designer guesses correctly, the help that appears at the user's initial request will solve the user's problem and halt the search for information. Of course, guessing what help is desired is a difficult process and often requires that help developers observe user problems during usability testing of the interface.

In some cases, we have recommended developing help text directly in response to information queries during testing: The users ask for assistance, the help developer responds with the minimal information necessary to move the user along—that text becomes the source of the help system.

Embedded help, a system more closely related to wizards and coaches, provides performance support in the context of the application. Since the help is always present, the user does not have to locate the information they need to complete a task. The information is provided to them immediately in the context of task performance. Development tools are now available to embed help within the application screens and to highlight the appropriate step in a process as the user proceeds through the activity. (For more information, see the HelpXtender from Wextech Software.)

Well-designed help systems and user guides should be as brief as possible to encourage the user to engage with the software application and apply their previous knowledge and experience to the task. However, it is especially important that the text in the interface, the information on the screen, be closely linked to the information in the instructional material. This approach to interface design recognizes that the help system and the user guides are integral parts of the interface and must be designed in association with the interface design.

For an interface to comply with the fourth minimalist principle—to support reading to do, study, and locate—means the following:

- Provide minimalist help and instructional text in user guides to support the user's task performance.

- Develop context-sensitive and embedded help systems so that the transition between actions in the interface and supporting information is as effortless for the user as possible.

- Carefully integrate the development of all aspects of the interface, including the development of help and additional instructional text so that all elements work together to support the user.

Summary

Minimalism provides us with insight into the design of user-centered interfaces. It deemphasizes the construction of pleasing screen layouts, placing emphasis more properly on user-goal-oriented design. Minimalism suggests, in some ways, that interfaces need to be designed to provide more information to the user than they do today. That information might come in the form of metaphoric interface structures, assistance in moving the user from their goals through the specific tasks that support the goals. It might come in the form of a richer textual content in screen design and certainly with a focus on selecting the right words and phrases to ensure that the user makes intelligent choices.

Minimalism also points to the need for close integration among the graphic interface, the help systems, and other forms of instruction. Close integration has been proved successful in the development of performance assistants such as wizards and coaches, which provide more text and more task direction than the typical menu-based design.

Finally, minimalism directs us to be aware of the real goals of our user and to explicitly support the achievement of those goals. Minimalism suggests that we anchor the tool in the task domain, rather than in the database. Such a perspective means that we must spend time understanding the *user* rather than designing screens.

Bibliography

Carroll, J.M. 1990. *The Nurnberg Funnel: Designing Minimalist Instruction for Practical Computer Skill.* Cambridge, MA: MIT Press.

Carroll, J.M. (ed.), 1998. *Minimalism Beyond the Nurnberg Funnel.* Cambridge, MA: MIT Press.

Lewis, C. and D.A. Norman. 1986. "Designing for Error." *User Centered System Design: New Perspectives on Human-Computer Interaction,* pp. 411–432. D.A. Norman and S.W. Draper (eds.) Hillsdale, NJ: Erlbaum.

Redish, J.C. 1988. "Reading to Learn to Do." *Technical Writing Teacher* 15, pp. 223–233.

Usability Testing Methods: Subjective Measures— Measuring Attitudes and Opinions

Joe Dumas, American Institutes for Research

This is Part II of an article on the subjective measures that are almost always used in some way in usability testing, such as pre-test background questions or post-test questions to assess participants' opinions of the usability of the product they have just used. In Part I, I described some of the research and practical advice in the literature about how to create effective questions while avoiding some of the sources of distortions in poorly designed questions.

Likert rating scales

Likert (1932) wanted to develop a method to measure attitudes that would be easy for respondents to fill out and for administrators to score. His method was to write statements and ask people to indicate their agreement or disagreement with them. For example:

Common Ground is a quality newsletter.

___ Strongly agree

___ Agree

___ Neither agree nor disagree

___ Disagree

___ Strongly disagree

Ratings with this general format became known as "Likert scales." Likert's method included creating several rating questions that he believed measured the same subjective state. The levels of agreement to the statements were then added or averaged by assigning the numbers 1 to 5 to each of the levels of agreement.

Researchers have since studied the validity and reliability of such scales as well as their level of measurement. (See the box on Scales of Measurement that follows.) Notice that in Likert's original work, respondents rated only their level of agreement (there were five levels), and the respondents never saw the numbers that researchers assigned to the levels of agreement.

Since Likert's original work, there have been many variations to the method. They include having scales that are different from agreement/disagreement, having scales that have fewer or greater than five levels, and having numbers assigned to the levels that respondents circle. The following statement is a distant cousin of a classic Likert scale:

Rate the quality of the newsletter:

Common Ground

Low   High

1   2   3   4   5   6   7

There is a large volume of literature on the format of Likert scale ratings, and it is relevant to usability testing. Let's begin by looking at the number of levels in the scale. As you can see, Likert's original scale had five levels, while the question above has seven. The literature does not favor any one number of levels, but rather specifies a range. After a rather exhaustive review of the relevant research, Cox (1980) recommends that there be somewhere between five and nine levels in a scale. Even though this range happens to be within Miller's (1956) "magical number of seven, plus or minus two" the similarity is purely coincidental. The range is not limited by short-term memory limitations.

To understand why this 5-to-9 range makes sense, imagine a rating scale with only three alternatives, such as "very easy . . . neither easy nor difficult . . . very difficult." There will be many respondents who'll choose the middle alternative even though they are leaning toward one end of the scale or the other. If we add two more alternatives on either side of the neutral point, such as "easy" and "difficult," some of the people who choose the neutral alternative would move to one of the new alternatives. As we add levels, we still gain useful information until there are more than nine or ten. According to Cox, there is little to be gained by adding more levels than nine. Having a 100-level scale does not necessarily give you more information than having fewer than ten levels. In fact, more than nine or ten alternatives may confuse some respondents.

Other researchers have looked at the value of having an odd number of levels so that there is always a neutral point, such as one labeled "neither easy or difficult." Some practitioners who avoid a neutral point have argued that it attracts those who lean toward one or the other polar alternative, though with little intensity. Consequently, eliminating a neutral level provides a better measure of the intensity of participants' attitudes or opinions. Some of those who do include a middle position assume that at least some respondents who choose it do have a neutral position and eliminating it increases the error of measurement of the scale. In a probing research study, Presser and Schuman (1980) showed that when they provided a middle, neutral level, 10 to 20 percent of the respondents chose it.

Surprisingly, very few of them would have chosen "don't know" if it were an option. Almost all of the respondents who choose the middle position have a neutral position or come from the other, nearby levels. The advice that Presser and Schuman give is that if you want to measure only the intensity of conviction, do not include a neutral position, otherwise include one. In our usability tests scales, we always use an odd number of levels and give participants the option of taking a neutral position. Of course, the most useful information about a rating is often what participants say about why they are selecting it.

What about the words you used to label the levels? They, too, make a difference. If I construct a scale with end points labeled "very difficult" to "very easy," have I constructed a scale with a single dimension, or are there two dimensions—a difficulty scale and an ease scale? Is saying that a product is "difficult" to use the same as saying it is "not at all easy" to use? The answer is probably not. Here is what Ostram and Gannon (1996) say about this issue:

"The categorical model of rating scale responses is based on the idea that the language of response options affects information processing in answering survey questions. It posits that the semantic end labels on rating scales activate corresponding cognitive categories. For example, let us assume that the researcher needs to measure perceptions of political candidates on an honest-to-dishonest continuum. If a rating scale is labeled from "not at all honest" to "very honest," the category honest is activated. But if the same question is asked on a scale from "not at all dishonest" to "very dishonest," the category dishonest is activated. Thus the category dishonest is not merely the inverse of the category honest; it is also a different knowledge structure. . . . A slightly more subtle issue is whether the end label "not at all honest" activates the same knowledge structure as the more directive label "dishonest."

The research investigating this issue is complex. One of the research methods that is used is called "exemplar generation," in which respondents are given rating point labels such as "honest" and asked to generate examples of behaviors that the name describes. Researchers can then examine the content of the examples generated by each name, as well as other measures such as how many examples are generated per name, the time it takes respondents to generate each example, and the rating by each respondent of how easy or difficult it was to generate the example. The primary issue of interest in these studies is how many cognitive categories are activated in giving meaning to any scale. The research shows that some rating scales activate one cognitive category, while others activate two or even three categories. The labels "honest" and "dishonest" activate different cognitive structures. The scale is clearly bipolar. Substituting "not at all honest" for "dishonest" makes the scale less bipolar, but the truly unipolar pattern is not activated until the end points "honest" and "very slightly honest" are used.

What this means for rating scales used in a usability evaluation process is that we need to be careful about the labels we use, especially when the labels appear to be bipolar. This is only speculation on my part, but the observation that usability test participants often give more positive ratings to products than their other measures would predict might be due in part to the use of end labels such as "easy" and "difficult." If "difficult" activates a different cognitive structure than "easy," participants may have good reason for staying within the one category "easy."

My anecdotal observation that participants frequently give very positive ratings to user interfaces with which they struggle leads me to believe that participants (1) are looking for cues to help them decide what is expected of them in a testing environment, and (2) want to be viewed as positive people. After all, they are usually compensated for their time and they believe, in spite of what we tell them, that we as testers are involved in some way with the product development. They are usually correct, though we try to remain neutral. (Whether we are ever really neutral is a question worth considering.)

Until we get some research on this issue, it seems likely to me that in a usability test, a scale that uses "very difficult" would activate a different cognitive structure than one that uses "not at all easy," or the more awkward "very slightly easy." Until research demonstrates otherwise, I suggest that usability testers stay with any format they have used over time, especially if it supports baseline data that will be used to assess new products or innovations. If testers have a choice of new formats, I suggest using end-point labels such as "easy" to "not at all easy" rather than "easy" to "difficult." We won't know whether these options yield different response patterns in a usability evaluation context until there is some quality research on this issue.

Finally, what about the numbers that you use in a rating scale? Consider the following two scales:

Not at all easy						Easy
1	2	3	4	5	6	7

Not at all easy						Easy
-3	-2	-1	0	+1	+2	+3

There is a formal similarity in these scale numbers. In the past, I have used the -3-to-$+3$ scale at times because it appears to give the participant a clear choice between the difficult and the easy end of the scale. When a user interface is poorly designed, I believed the -3-to-$+3$ scale gives participants a kind of permission to choose a negative rating.

But the research literature suggests that the -3-to-$+3$ scale may actually push participants more toward the positive end of the scale than a 1-to-7 scale (Schwarz, et al, 1991). This research suggests that a respondent is less likely to make a rating of -1 on a -3-to-$+3$ scale than they are a 3 on a 1-to-7 scale. There have been several well-planned research studies showing that respondents who have the −3 to +3 scale give significantly more positive ratings than respondents who rate the same quality using the 1-to-7 scale. This finding is consistent both when the respondent is doing a rating that is self-revealing, such as "rate how successful you have been in life," and when the respondent is doing a rating that is relatively neutral, such as rating how satisfied a described person is with his or her own health. In our lab, we no longer use the −3-to-+3 scale.

The semantic differential

Another question format with a long history of success is the semantic differential. It was first described by Osgood, et al, (1957). They were interested in understanding the connotative meaning people attached to objects or events. They asked respondents to rate objects such as people or events on a number of bipolar adjective scales, such as:

Good ____ ____ ____ ____ ____ ____ ____ Bad

Strong ___ ____ ____ ____ ____ ____ ____ Weak

Active ___ ____ ____ ____ ____ ____ ____ Passive

Respondents put a check on one of the seven lines between each pair of adjectives.

Over the years, many studies have been done with the semantic differential on people from over 25 countries (Dawes, 1972). In these studies, people have been asked to rate many objects with a large number of bipolar adjectives. The data is quite consistent. The responses tend to cluster into three dimensions: an evaluative dimension, typified by "good–bad"; a potency dimension, typified by "strong–weak"; and an activity dimension, typified by "active–passive."

The evaluative dimension usually is the most relevant to usability testing. Keep in mind, however, that because the reliability of any one pair is relatively low, you usually use three or four adjective pairs to assess one dimension. You then average the values on the pairs to obtain one score for each individual on a dimension. For the evaluative dimension, you could use pairs such as favorable–unfavorable, pleasant–unpleasant, and attractive–unattractive, in addition to good–bad. You could ask test participants to rate a user interface on these adjective pairs and use the average rating as a measure of the participants' evaluation of usability. See Alreck & Settle (1995) for practical guidance on how to use a semantic differential scale.

Scales of measurement

Throughout the history of psychological measurement, researchers and practitioners have put a lot of work into demonstrating that psychological measures have ordinal and interval properties. Here's why.

Measurement is the assignment of numbers to objects in a systematic manner as a means of representing properties of the objects (Allen & Yen, 1985). There are four levels of measurement that are relevant to subjective measures:

Nominal level — These numbers represent only the *distinctiveness* or *identity* of objects. The numbers on the back of basketball players are nominal. They are just labels. Every player has a different number, but the numbers themselves don't measure anything but identity. The number a player has, such as "10," does not measure his or her ability to play basketball. The labels might just as easily be letters of the alphabet or just the players' names.

Ordinal level — These numbers represent the *ordering* or *magnitude* of a property of an object. If product A is more usable than product B, we want a measure of usability for product A to be larger than a measure for product B. At the ordinal level, the size of the difference between A and B is not relevant, only the order of the measures. When we ask participants to rank order products on ease of use, we are using an ordinal measure.

Interval level — These numbers represent *equal intervals in the magnitude of a property.* Interval scales measure not only that Product A is more usable than Product B, but by how much. When a measure at an interval level assigns a 10 to Product A and a 5 to Product B, Product A is not only more usable on the measure than Product B, but is five units more usable. The temperature scale in Fahrenheit is an example of an interval scale. When we use rating scales (see body of this article), rather than rankings, we usually assume that we are using an interval level of measurement.

Ratio level — These numbers have *interval properties, but also have an absolute zero*. In ratio-level measures it is possible to have zero or none of a property. If we had such a measure in usability measurement, we would be able to measure when a product has zero usability. Few, if any, attitude measures have this property.

When you have a scale that measures at the interval level, you can use it to assess not only that A is superior to B, but by how much.

So what does all this mean to those of us who do testing on a day-to-day basis? You can see that coming up with an ideal question or even a question format takes some careful analysis and a willingness to continually assess whether the questions you use are really getting at what you want to measure. I have four thoughts for usability practitioners regarding using questions as subjective measures:

1. Always consider *why* you are asking a question and how participants might be interpreting the words and the format you use.

2. When you find a question that works for you, stick with it. Using the same question in many tests gives you some context with which to interpret answers to it.

3. Use questions as a way to stimulate discussion about the topic of interest. We have all had the experience of the participant who rates an interface easy to use, then, when prompted to elaborate, describes just the opposite.

4. Try never to restrict your evaluation of usability to any one measure, especially a subjective measure. When something is wrong with an interface, the problem manifests itself in several ways that together document the problem. Plan to collect several measures that allow you to triangulate on a problem rather than depending on one measure.

Additional thoughts

Usability test practitioners who have read these articles on subjective measures occasionally have asked for advice on using test questions from published questionnaires or from questionnaires that other testers have already "debugged." My advice is as follows.

Published questionnaires

There are a wide variety of questionnaires that other people have developed and published. Keep in mind that some of these questionnaires were developed to be substitutes for a usability test—not to be used within a test. Here are three that are commonly cited:

Questionnaire for User Interface Satisfaction (QUIS)

This questionnaire has been used by many evaluators over the past ten years—in part, due to its accessibility. It is published in Shneiderman's book, the 1992 second edition and the 1997 third edition. In concept, it has a set of general questions for an overall assessment of a product, then a set of detailed questions about HCI components. The detailed questions cover the following user-interface components:

- Screen design
- Terminology and system information
- Learning to use a product
- System capabilities

Because these categories are not always relevant to every product, testers often select a subset of the questions or only the general questions. There is a long form of QUIS (71 questions), and a short form (26 questions). Each question uses a 9-point rating scale, with the end points labeled with adjectives. Here is an example:

Characters on the screen are:

Hard to read   Easy to read
  1   2   3   4   5   6   7   8   9

The software usability scale (SUS)

SUS was created to be administered quickly. Hence it lends itself to the debriefing section of a usability test. It was created by a group of professionals at what you may remember as Digital Equipment Corporation (Brooke, 1996).

The ten SUS questions have a Likert-scale format: A statement followed by a five-level agreement scale. Here is an example:

I think that I would like to use this system frequently:

Strongly   Strongly

Disagree   Agree

| 1 | 2 | 3 | 4 | 5 | 6 | 7 | 8 | 9 |

The strength of SUS is that it is short and, consequently, fits nicely into the time scale of a typical usability test.

The Software Usability Measurement Inventory (SUMI)

SUMI is really a usability evaluation method of its own (Kirakowski, 1996). SUMI has been subjected to a series of validity and reliability studies by its developers. It was created only to evaluate software. It is a well-constructed instrument with 50 questions that break out into six subscales:

- Global

- Efficiency

- Effect

- Helpfulness

- Control

- Learnability

The Global subscale is similar to QUIS' general questions. The developers have created norms for the subscales so that you can compare the software product you are evaluating with the answers other people have given for other products. For example, you could show that the product you are evaluating scored higher on all of the subscales than the average product evaluated by SUMI.

The questionnaire comes with a manual for scoring the questions and using the norms. The developers recommend that the test be scored by a trained psychometrician. If you know one, make friends.

References

Brooke, J., (1996), "SUS: A 'Quick and Dirty' Usability Scale," in Jordan, P., Thomas, B., Weerdmeester, B., & McClelland, I. (eds.) *Usability Evaluation in Industry.* London: Taylor & Francis, pp. 189–194.

Kirakowski, J., (1996), "The Software Usability Measurement Inventory (SUMI): background and usage," in Jordan, P., Thomas, B., Weerdmeester, B., & McClelland, I. (eds.) *Usability Evaluation in Industry.* London: Taylor & Francis, pp. 169–177.

Shneiderman, Ben, (1997), *Designing the user interface: Strategies for Effective Human Computer Interaction.* Reading, MA: Addison–Wesley, 3rd Edition.

Usability Testing Methods: Think-aloud Protocols

Joe Dumas,
American Institutes
for Research

Watching a test participant think aloud is probably the signature quality of a usability test. Wherever a test is held around the world, there will be a test participant talking as he or she works. This talking is one of the factors that gives usability testing so much credibility as an evaluation tool. Visitors often are stunned by the realism of the experience. But what are we really hearing when a participant "thinks aloud," and how do we get the type of "think aloud" that is most helpful to us?

What do we want participants to do when they think aloud in a usability test?

Answer: More than just play-by-play, also why, expectations, reactions, etc. In a usability test, the participants need to know that the focus of thinking aloud is on their interaction with the product and on all of their inner experiences as they interact. If we restricted participants only to thoughts, as in some cognitive psychology research, we would get only an uninformative play-by-play description of what participants read and what actions they perform.

Is this what was originally meant by "thinking aloud"?

Answer: Not always. Researchers separate three levels of thinking aloud.

The technique of thinking aloud has been part of psychological research since about 1900. (See Ericsson & Simon, 1993, for a thorough review of the research and issues.) The method survived the criticisms of the early behaviorists, lead by John B. Watson, who said psychologists should restrict the data they collect to observable phenomena and that what a human says and does are independent, that is, have no clear relationship to one another.

With the rise of cognitive psychology over the past 30 years, thinking aloud has again become a standard method for studying human thought. The conditions under which it provides access to silent thoughts are now believed to be well understood. But the focus of this research is on thoughts, not expectations, or emotions, just thoughts.

Level 1 verbalization: Just thinking (no explanations)

Researchers believe that there are three levels of thinking aloud. Level 1 verbalization occurs when participants are instructed to say:

. . . what you think about when you find answers to some questions that I am going to ask you to answer. In order to do this, I am going to ask you to think aloud as you work on the problem given. What I mean by "think aloud" is that I want you to tell me everything you are thinking from the time you first see the question until you give your answer. I would like you to talk out loud constantly from the time I present each problem until you have your final answer to the question. I don't want you to try to plan out what you say or try to explain to me what you are saying. Just act as if you are alone in the room speaking to yourself. It is most important that you keep talking. (Emphasis added, Ericsson & Simon, 1993, p. 378.)

Notice that the emphasis is on thoughts and that the participant is told not to plan or explain what he or she is thinking. The participants are then given several practice problems to make sure that they are following instructions. The practice problems are most often thought problems, such as "multiply 24 by 36 in your head" or "count the number of windows in your house." The participants are given practice problems until they are giving only Level 1 verbalizations.

In addition to the instructions to the participant, the person conducting the research is suppose to sit behind and out of view of the participant and also not to say anything to the participant unless the participant doesn't follow the instructions. The researcher's role is not to engage the test participant in social dialog, but to record what happens and to make sure the participant follows the instructions. If participants stop talking, they are prompted to "keep talking." If the participant reports more than what he or she is thinking, such as by saying why they are doing a task, they are instructed again to "just say out loud what you are thinking as you solve the problem."

In these studies, there is a think-aloud condition and a silent condition in which a different group of participants are given the same problems, but asked to work without speaking. After several decades of research with this technique, the conclusion is that the above instructions and practice allow participants to say out loud what they are thinking. Research test participants are assumed to be saying out loud what they have in their short-term memory. There are many strategies that people could use to solve a thought problem. But the research shows that the way participants solve the problem and what they say out loud match each other when they are giving Level 1 verbalizations. If the inner thoughts and what participants say out loud were not related, why would think-aloud protocols match what participants do in these studies? In some, but not most, studies, it takes somewhat longer to solve the problem when participants think out loud, but there are no other differences between the silent and think-aloud groups.

Level 2 verbalization: Same but when manipulating non-verbal information

Level 2 verbalizations are similar to Level 1 in terms of the instructions participants get, but in these studies, the problem requires the manipulation of non-verbal information, such as geometric shapes. For these problems, the participant has to code the results of the problem solution into words to say them out loud. Other than this coding of the results of the problem, verbalizations while solving a non-verbal problem are assumed to mimic internal thought.

Level 3 verbalization: Thinking plus explanations

Level 3 verbalizations are qualitatively different from the other levels. In addition to the standard instructions, participants in Level 3 verbalizations are given supplemental instructions, often to "explain each step as thoroughly as you can" or "say not only what you are thinking, but why." Under Level 3 verbalization instructions, the assumption is that the supplemental instructions change the inner processes that participants use. We are no longer getting a record of what participants read out of their short-term memory. Rather, we are getting the participants' interpretation of the process they are using or the reasons they have selected a strategy. The assumption that the participants are saying aloud only what they are thinking to themselves is no longer justified. In such situations, the relationship between a think-aloud and a silent condition is no longer one-to-one. In these studies, there are many differences between the think-aloud and the silent groups. The researchers who study this technique tell us that what we are getting from the participants' Level 3 verbalization is more than inner thoughts.

This research in cognitive psychology shows that research test participants do report out loud what they are thinking as they do a task, but only when they are properly instructed and practiced. If participants are asked to report other internal states, what they report is a construction of what they are thinking and other internal states they may be experiencing.

What happens in usability studies with different instructions about thinking aloud?

Answer: Let's look at three research studies that can help us answer that question.

The value of Level 3 verbalization

In an early article, Denning, et al, (1990), reported on the use of thinking aloud in usability tests in their organization. The article describes how "think aloud" is done and how videotapes of think aloud are used in data analysis. The authors do not quote their think-aloud instructions in the article, but do say, "Subjects are instructed to verbalize *all* their thoughts while they use functional parts of the product and to verbalize their *expectations* when they tried to use something that was obviously unimplemented" (emphasis in the original). It seems clear that usability test participants were given instructions that produce Level 3 verbalizations, that is, are not restricted to thoughts. In the article, the authors also say, "The video medium is certainly powerful in its own right, but its real advantage for us is that it allows us to capture vivid examples of users' thinking aloud, verbalizing their frustrations and successes with the product."

Level 1 verbalization can lead to "play by play" descriptions

Bowers and Snyder (1990) conducted a study to compare the advantages and disadvantages of having test participants think out loud as they work, called "concurrent" think aloud, with thinking out loud after the session is over, called "retrospective" think aloud. In the retrospective condition, the participants performed tasks in silence, then watched a videotape of the session while they thought aloud. This is a very interesting study because of its implications for usability testing.

The group of participants who performed concurrent thinking aloud were not given typical think-aloud instructions for a usability test. Instead, their instructions were typical of a think-aloud research study. They were told to "describe aloud what they are doing and thinking." They were not told to report any other internal experiences. In addition, they were never interrupted during a task. There was no probing. Any encouragement they needed to keep talking was done only between tasks. The retrospective participants were told that they would be watching the videotape of the session after the tasks and would be asked to think out loud then.

There were several interesting results. First, there were no differences between the concurrent and retrospective groups in task performance or in task difficulty ratings. The thinking out loud during the session did not cause the concurrent group to take more time to complete tasks, to complete fewer tasks, or to rate tasks as more difficult in comparison with the performance of the retrospective participants. This finding is consistent with the results from the other think-aloud research studies.

The differences between the groups were in the types of statements the participants made when they thought out loud. The concurrent group verbalized about four times as many statements as the retrospective group, but the statements were almost all descriptions of what the participants were doing or reading from the screen. The participants who did concurrent thinking aloud were doing exactly as they were instructed; they were attending to the tasks and verbalizing a kind of "play by play" of what they were doing. The participants in the retrospective condition made only about one-fourth as many statements while watching the tape, but many more of the statements were explanations of what they had been doing or comments on the user-interface design. "The retrospective subjects . . . can give their full attention to the verbalizations and in doing so, give richer information."

This study shows us what would happen if we tried to get participants in usability tests to report only Level 1 verbalizations and did no probing of what they were doing and thinking. Their verbalizations would be much less informative. The study does show that retrospective thinking aloud yields more diagnostic verbalizations, but it takes 80 percent longer to have the participants do the tasks silently, then think out loud as they watch the tape.

Thinking aloud (Level 3) helps uncover problems

A study by Virzi, et al, (1993), compared the value of silent and think-aloud conditions at uncovering usability problems. While the authors do not quote their think-aloud instructions, the instructions appear to be typical for usability testing in asking participants to report all of their inner experiences out loud that relate to interacting with the product. The silent group performed tasks without verbalization. In this study, the authors do not report whether there were any performance differences between the silent and think-aloud conditions. After the test was run, the three authors, all experienced usability specialists, listed the problems that participants who thought out loud uncovered. They then looked at the performance data of the silent group for evidence of usability problems. They mention two types of data that revealed problems: high variability in task performance and the inability of participants to complete tasks. Notice that the authors did not impose their knowledge of the usability problems with the product onto the data from the test. The results show that, without think-aloud or the authors' imposing their knowledge of usability problems, the silent condition was less fruitful at uncovering problems. The silent condition uncovered 46 percent of the problems with the product, while the think-aloud condition uncovered 69 percent. This study clearly shows the advantages of think-aloud protocols. It also suggests that in many cases, participants' think-aloud protocols provide evidence of usability problems that do not otherwise show up in the performance data.

So what does this all mean for usability testing practice?

If you have conducted any usability tests, you can see how different thinking aloud in usability tests is from thinking aloud in cognitive psychology research studies. But the think-aloud procedures used in research studies and the findings of those studies are still relevant to what we practice in usability testing. The research studies we just reviewed on thinking aloud in usability test situations show the value of thinking aloud. Without it we would uncover fewer usability problems and waste a valuable resource. But the way we instruct usability test participants leaves them free to decide what they report.

How do we get participants to do the right type of think-aloud for a usability test?

Answer: Give instructions that explicitly ask for the types of information we want. Demonstrate it to participants. Have them practice.

In usability tests that I administer, I demonstrate thinking aloud while I use a product, then have participants practice thinking aloud. For example, I demonstrate thinking aloud as I "find out if there are enough staples in this stapler." Then I ask the participant to "think aloud as you take the ink cartridge out of this pen." This demonstration puts the focus on the interaction with a product. (For further discussion of instructions, see Dumas & Redish, 1999.)

In usability tests, the instructions or demonstrations given to participants ask them to report more than thoughts. They typically are asked to report all of the experiences they have in their interaction with the product, including such states as expectations and feelings. For example, in the demonstration of opening and taking the staples out of the stapler, the test administrator reports other experiences, such as "that was a little more difficult to do than I expected" or "I like the fact that it says 'lift cap to load' on the stapler." In response to a participant who, while practicing, reports more than a thought, the administrator might also say, "That's right, I want you to say more than just what you are doing."

By using these think-aloud instructions in usability testing, we are leaving it up to the participants to decide what to report and how to interpret their inner states. When a participant says, "I am really frustrated by this product," we leave it to him or her to determine what "frustration" means to them. When two participants use the same word, we do not know whether their internal states are the same. Is one participant's "frustration" the same as another's? We don't know. That may be one reason why we combine several measures in our data analysis. We report what a participant said, their task time, and their errors. But the Virzi, et al, study shows that there are many problems that don't show up in the performance data, only in the think-aloud protocol. So if two test participants express the same feeling, let's say frustration, but there are no other indicators of a usability problem, are we justified in saying there is a problem? Are we as testers easily convinced when a participant's statement agrees with our own evaluation of a problem? This question about the validity of think-alouds as indicators of problems deserves a longer discussion in another article.

How does participants' thinking aloud affect the test administrator's role?

Answer: potentially creates conflict between the test administrator's two roles.

Test administrators usually do not provide any additional instructions about thinking aloud as the test session proceeds, except when the participants stop talking. A common prompt to participants who stop talking is, "Tell me what you are thinking." In the think-aloud research, such friendly, social prompts are discouraged in favor the short and to-the-point, "Keep talking."

In a diagnostic usability test, the test administrator is performing two roles: (1) being friendly and encouraging to help participants feel comfortable and to keep the focus of the test on the product, not on the participant, and (2) being an unbiased observer and reporter of what the participant says and does. In the role of unbiased observer, test administrators by their actions should not change the likelihood that the participant will report positive or negative thoughts or feelings toward the product.

The friendly role and the neutral observer role come into conflict when participants make a strong statement expressing an emotion, such as, "I hate this program!" Almost anything the test administrator says at that point can influence whether the participants will report more or fewer of these negative feelings. Here is a list of possible responses by a test administrator:

"Tell me more about that" — relatively neutral in content, but could be interpreted as encouraging more negative statements.

"That's great feedback" — again, relatively neutral to someone who has training in test administration, but, I believe, sounds evasive to participants.

"Those are the kinds of statements that really help us to understand how to improve the product" — reinforcing the negative.

"I really appreciate your effort to help us today" — says nothing about the content of what they said, and is part of playing the friendly role with participants. Will the participant hear it that way?

" . . . " — silence. Neutral in content, but how will it be interpreted? In human interaction, one person's silence after another's strong statement is almost always interpreted as disagreement or disapproval. Without any other instructions, the participant is left to interpret the test administrator's silence—you don't care, you don't want negative comments, strong feelings are inappropriate in this kind of test, etc.

By the way, not all of the biasing responses to emotional statements are verbal. If you are in the room with the participant and you take notes when participants make an emotional statement, you may be reinforcing them to make more.

Any of these responses could push the test participant to utter more or fewer strong feelings. The only way I can see to avoid this conflict in roles, and thereby avoid influencing participants or making them feel that we are strange people for not reacting to their comments, is to tell them what our roles are in the test situation. We normally tell participants that one of our roles is to work with them as co-evaluators and that we are not testing them, but the product. That is the friendly role. But as part of our instructions, we also could say:

"I am going to be working with you today, but my role is to communicate what you say and do to the product developers. I do not want to bias you toward liking or disliking the product. So don't be surprised if sometimes I don't say anything in response to your positive or negative comments or if my response is something neutral such as, 'That's great feedback.' It is not that I don't care what you say or that I didn't hear you. I am simply maintaining my neutrality so that I don't bias you one way or the other. I am listening to what you say. Feel free to say what you want and be sure that I will communicate it to the people who make this product."

You may need to remind the participant during the session of your role as unbiased observer, especially if the product is difficult to use.

I think that being clear and straightforward with our test participants about our roles will help them interpret our actions as we intend them and will make it less likely that we will influence what they say.

One more point: the conflict in the roles changes with the type of test. We have been talking about the conflict in a diagnostic test. But in a comparison test, the role of neutral observer becomes even more important. The test administrator can't bias the participant toward any of the products. This constraint makes it more difficult to be friendly and helpful in a comparison test.

Returning to the question we started with, I think you will agree that when participants think aloud in a usability test, there is a lot going on. The notion that we are hearing an unedited record of inner thoughts is too simple. The research shows that what participants say as they think aloud is an invaluable aid to diagnosing usability problems. But what they say also is influenced by their interpretation of our role, our instructions to them, their interpretations of those instructions, and by our reactions to their reactions. We need to keep our potentially conflicting roles as friendly co-evaluators and unbiased observers clear to both our test participants and ourselves.

References

Bowers, V., & Snyder, H., (1990). "Concurrent versus retrospective verbal protocols for comparing window usability." Proceedings of the Human Factors Society 34th Annual Meeting, pp. 1,270–1,274.

Denning, S., Hoeim, D., Simpson, M., & Sullivan, K., (1990). "The value of thinking-aloud protocols in industry: A case study at Microsoft Corporation." Proceedings of the Human Factors Society 34th Annual Meeting, pp. 1,285–1,289.

Dumas, J.S. & Redish, J.C., (1999). *A Practical Guide to Usability Testing*. Exeter, UK: Intellect Books.

Ericsson, K.A. & Simon, H.A., (1993). *Protocol Analysis: Verbal Reports as Data*. Cambridge, MA: MIT Press.

Virzi, R.A., Sorce, J.F., & Herbert, L.B., (1993). "A comparison of three usability evaluation methods: Heuristic, think-aloud, and performance testing." Proceedings of the Human Factors and Ergonomics Society

37th Annual Meeting, pp. 309–313.

Wright, R.B. & Converse, S.A., (1992). "Method bias and concurrent verbal protocol in software usability testing." Proceedings of the Human Factors Society 36th Annual Meeting, pp. 1,220–1,224.

Consumer "In-Home" Usability Testing

David Mitropoulos–Rundus,
Compuware Corporation
and Jerry Muszak,
Eastman Kodak

Introduction

Observing product usage in the intended environment provides "real world" data that a usability test in the lab may not. Given the continual introduction of new and improved products intended for leisure, utility, and home-office use, there is an ever-increasing need to evaluate the performance of these products in their natural environment. We have designed "in-home" evaluations for a broad range of existing products, evolving products, and new product concepts for which functional prototypes have been created, and we have conducted evaluations in over a hundred homes. Based on this experience, we have evolved criteria for determining when in-home testing is appropriate, and for creating practical evaluation protocols. The intent of this chapter is to provide you with enough insight and direction so that you can make better decisions on whether and how to conduct such testing.

This chapter is divided into three sections. Section I defines in-home testing, compares it to other types of field and lab testing methods, and describes the data collected. Section II reviews the feasibility for conducting "in-home" testing and presents criteria to assist you in determining whether in-home testing is most appropriate for your product and situation. Section III steps you through design and implementation of an "in-home" usability test.

I. What is "in-home" testing?

This section defines in-home testing, compares it to other test methods, and describes the data collected. Consumer in-home usability testing is one way to collect data on product perceptions, usage, usability, and durability. In this method, participants have the product in their home for several days to several weeks. At the conclusion, a test team visits the home to collect and review data on usage and product perception.

Similarities to field testing

Consumer in-home usability testing is a special case of field testing. As with typical field testing, less control is traded for more reality. The challenges these methods share include:

- Product robustness: The product must be able to sustain transportation and varied usage.

- Logistics: Appropriate participants must be located within a reasonable geographic area.

- Lack of control: Participants cannot be given, or be expected to follow, a strict protocol, and some data-collection occurs without direct observation.

Comparison to laboratory usability testing

Below are some comparisons between in-home and lab usability testing.

Testing in the usability lab	Testing in the home
Unnatural/unfamiliar environment to participant	Natural/familiar environment to participant
Natural/familiar environment to experimenter	Unnatural/unfamiliar environment to experimenter
Getting the participants to the lab	Getting the lab to the participants
Experimenter in control	Participant in control
Features of interest to the experimenter tested completely and systematically	Only features of interest to participants tested
Data on many usability problems	Data on some usability problems
Few or no interruptions (quiet, controlled setting)	Interruptions occur during home visit/interview (dogs, cats, children, phone ringing...)
Difficulty distinguishing, i.e., "short-term" from "long-term" usability problems (they are all just "usability problems")	Could miss some "short-term" learning" usability problems that may be resolved prior to the home visit/interview

The unique value of in-home usability testing

In-home testing can provide many types of information. As with laboratory usability testing, in-home usability testing can uncover usability issues which can be folded into product designs, redesigns, packaging, and user manuals. In addition, in-home usability testing may be worth the extra time, effort, and expense for the following reasons:

• Home usage environment data: Many aspects of the home environment (e.g., ambient room light, distance between components, viewing distance, concurrent activities, competing activities) cannot be discovered in the laboratory. This data may be valuable for the product under evaluation as well as related products, and it cannot be collected any other way.

• Realism and credibility of data: Because the product was used in the home for a period of time, people tend to believe the data more.

• Generates interest, excitement, and assurance with product team and management: People tend to listen and get excited when they hear about their products being tested in their intended environment.

• Better predictor of usage patterns: Participants have tried to use the product on their own, and resulting usage patterns may be more natural than those observed in a laboratory usability test. These product usage patterns can provide a more intuitive template for designing user instructions.

• More complete product perceptions: Participants have tried to use the product on their own and for a longer period than in a typical laboratory usability test, and resulting product perceptions may be more extensive and complete than those formed during a laboratory test. These product perceptions can include positive aspects that can be used in sales and promotion.

• Some control over data-collection: Although the product is not tested in the lab, the home visit at the conclusion of a study permits observation of the usage environment, demonstration of usage by participants, and administration of controlled usability protocols.

Unique aspects of "in-home" testing

Unique aspects and challenges of consumer "in-home" testing include the following:

- User attitude and behavioral differences: There is a tendency for leisure and relaxation over business. Also, you are on the participant's private property, which is quite different from the participant being in your usability lab. The environment is "owned" by the participant, and resulting behaviors may be quite different from business, production, and laboratory environments. We have experienced everything from "welcome to our humble home" to "businesslike" to "territorial" to "hostile."

- Getting participants to use the product/system: In the field-testing business environment, there is often a scheduled or business reason to use the product/system under evaluation. In a home environment, schedules are often looser. Some level of direction and encouragement is required to ensure usage, but too much direction can result in "instructed" rather than "typical" usage patterns.

- Family/friend participation considerations: Family members or friends may be present during the in-home visit. A decision needs to made up front regarding where to focus the data-collection. Whatever data is deemed important, everyone present needs to feel involved.

- Testing time: Typically, testing is limited to late afternoon/early evening, 5:00 to 8:30 p.m., Monday through Thursday. One must work around dinners and family activities.

- Anybody home? In the field-testing business environment, there are fixed hours of operation during which you can at least enter the facility. In the home environment, you have to wait outside if nobody is home, which can be a serious issue, especially when testing in Rochester in January.

- Safety/security: Given the variability in, and lack of familiarity with, the range of neighborhoods and homes that may be visited, "in-home" testing elevates the issue of safety and security when compared to visiting participants in a place of business or inviting them to your test lab. We encourage conducting each home visit with at least two people and carrying a cellular telephone.

- Allergies: Home environments are more variable and less controlled than most business environments. Test team members with allergy problems must therefore be aware of, and prepared to handle, the variety of pets, smoke, and dust that are encountered.

- Dogs in love: They can fall in love with *you*, too—watch your leg!

II. Determining if "in-home" testing is feasible for your product

This section presents criteria to assist you in determining whether in-home testing is most appropriate for your product and situation.

Is an in-home test feasible and appropriate for your product?

Equipment, logistics, and flexible schedules are key to conducting in-home testing successfully. Without careful planning and attention to a number of unique details, your in-home test can turn into an inefficient real-world test. Timelines can quickly extend to unreasonable completion times and costs can quickly add up.

The first step in planning a successful in-home test is to determine if it is the appropriate thing to do. We recommend that you answer as many questions as possible through lab testing and focus groups (we recommend these methods because they are typically less expensive and less time-consuming):

- Anticipated usage
- Feature usage/usability
- Unpack/setup usability

Next, determine if you have addressed key issues adequately. If there are still issues that are best addressed through in-home testing, determine if you can collect your data remotely or indirectly. Several options include:

- Telephone survey of current product owners.
- Focus groups or one-on-one interviews with current product owners.
- Product handouts for home usage followed up with interviews or focus groups.

If visiting the home is still high priority, you will also need to assess the following:

~ Is the product/system/prototype "robust enough" to withstand testing? Prototypes, engineering models, and pre-production models may not be able to withstand repeated or lengthy use. Alpha and beta software developed on one platform may have too many "bugs" to work efficiently enough for a test. Also, early versions of software may not have been tested on the range of platforms that will be used in the homes of the people you plan to test.

~ Do you have enough products/systems/prototypes? If you have a product for each participant plus a few extras (replacements for products that break or malfunction), you are in good shape. If you have fewer products than participants, however, the time required to collect all data will at least double. If you have only one or two products, in-home testing will most likely be impractical because of the time it would take to test a reasonable number of participants.

~ Do you have the time? The ratio of products to participants discussed above is most critical in determining if you have the time. It is also important to consider the period of home use (which often ranges from several days to several weeks), and scheduling and conducting a visit to each participant's home at the conclusion of the usage period. One home visit team can practically conduct about 8 home visits per week (an average of two visits between 5:00 and 8:30 p.m., Monday through Thursday).

~ Do you have the funding? In-home testing tends to be more expensive than testing in the lab. One should consider overtime payment for evening work hours, additional time spent traveling, and mileage reimbursement.

If you can answer yes to all of the above questions, then it is appropriate and reasonable to design and conduct your in-home product test—go for it!

III. Designing and conducting the "in-home" usability test

This section steps you through the design and implementation of an in-home usability test. If you have determined that in-home testing is appropriate for your product or situation, it is time to plan your test. The steps listed in this section progress in a logical order of events that will allow you to successfully plan and execute an in-home study. We hope it helps you to either successfully design and implement a practical in-home test the first time, or to improve your efficiency with successive in-home tests.

Step 1: Determining what data to collect

In-home usability testing is often more expensive and time-consuming than laboratory usability testing. Bearing this in mind, it does not make sense to conduct an in-home usability test to collect data that can be captured just as well through a laboratory test. Instead, an in-home test should be used to collect data that cannot be collected at all, collected as efficiently, or collected as well through laboratory testing.

Data that can be collected only through in-home testing. Home use environment data can be accurately captured only through in-home testing. This data includes a viewing distance (e.g., for televisions, for computer displays):

- Ambient room light

- Other equipment that interfaces with, or competes with, the product you are testing

- Identification, configuration, and integration of other products with the product you are testing (e.g., audio and video equipment, computers).

Data that is more efficiently captured through in-home testing. In-home testing is more efficient when a number of home environments must be sampled or simulated. A single home environment can be created in the lab, but simulating several different home environments is more difficult. Depending on the range of viewing environments, equipment components, and configurations required for your test, it may simply be less expensive and less time-consuming to collect this data by visiting homes. In addition, it may be difficult to even determine what the home environments are, much less simulate them, without first sampling them.

Data that is "more robust" if captured through in-home testing. If an important aspect of your test is successful hook-up and integration with other products in the home, an in-home test may give you "better" data. If the hook-ups are straightforward and common, you may be able to simulate a test well enough in the lab. If equipment and configurations are complex and vary widely, you may be better off letting the participants try with their own equipment. Having participants hook up your piece of "unfamiliar" equipment with their "familiar" equipment may be the only way to determine if this will occur successfully. If, on the other hand, you have participants in the lab hooking your piece of unfamiliar equipment to other unfamiliar equipment, it will not only be difficult to determine if the overall performance in the home will be successful or not, but you may also find it difficult to determine whether the source of the problems is the equipment you are testing or the supporting equipment.

One can often achieve longer periods of use with in-home testing when compared to scheduled lab testing. We also hope that product use will be more natural in the intended environment. As a result, participant data regarding preference, performance, utilization, and purchase interest are more credible and realistic because they are based on greater product experience and understanding. Finally, in-home testing creates a feeling of realism that lends credibility to the data, and tends to generate interest and excitement with the product team and management.

Other data-collection considerations. How many people will be your participants in a single home? An in-home test provides an opportunity to collect data from one person, or from an entire family. You will need to decide up front from whom you want to collect data, and structure your interview format and data-collection sheets accordingly. It is very easy to gather data from just one person at the home, and it is very difficult to gather complete data from everyone who may be in attendance during the in-home test. We recommend collecting complete data from one primary participant and collecting only comment data from others in attendance.

As a final consideration when selecting the data you plan to collect, design your study so that the home interview can be completed in one hour or less. As with many evaluations, participants tend to become fatigued or "less interested" after one hour of testing.

Step 2: Determining a schedule for study completion

In-home usability testing requires more time than laboratory usability testing. To create a realistic schedule and sample size, consider the following issues.

Number of products. From the time the product is given to a participant to the time the final home interview is concluded, one product is unavailable. This duration, combined with the time it takes to get the product to the next participant, will help determine a realistic turnaround time for your tests.

Reasonable period of home use. Our recommendation is to give participants either a relatively short period of home use or an extended period of home use. Following are our suggested home use periods:

One week including a weekend. Making sure your participants have the product over a weekend is particularly important because the weekend is usually the time when your product will be most heavily used and most heavily shared with family, relatives, and friends. The weekdays provide time for learning and also provide use time in case the participant is out of town for the weekend. One would intuitively expect a longer duration to provide better use data (i.e., two weeks are better than one). The incremental value of home use for two or three weeks compared to one week, however, has not been significant for the products we have tested. We have found significant added value in providing only much longer periods of home use (described below).

Two or more months along with other incentives (product giveaway, option to purchase at a reduced cost). This duration, along with options for the participant to have the product permanently, seem to elicit patterns of use that may be more representative of actual use by product owners.

Limited hours for data collection. Data collection in the home tends to be limited to late afternoon/early evening, 5:00 to 8:30 p.m. If you are testing your company's or your client's employees, however, you may find that a number of them can justify using their work time to participate in your study. As a result, you may be able to test more of your participants during normal business hours, thereby increasing the number of participants you can test in a day. Still, you should plan for an average of two home visits per day.

Limited days of the week for data-collection. In addition to the limited testing hours, data-collection tends to be limited to Monday through Thursday. Friday through Sunday are considered weekend nights, and participants are very reluctant to schedule testing during these times. If you have one dedicated test team, these constraints (i.e., two homes per day, four days a week) limit your testing to a maximum of eight homes in a week.

Step 3: Putting together an in-home interview team

Two experimenters are the ideal number for performing each in-home test.

Safety and security: Two people provide security and safety for the range of neighborhoods and homes that may be visited for testing.

Fast and efficient data-collection: Two experimenters can often collect data concurrently (e.g., one experimenter records home environment data while the other experimenter interviews the participant). This will minimize the time spent in each home, which will help ensure that you do not overstay your welcome and will also increase your ability to conduct more than one in-home test that day.

Two is not too many: More than two people can cause the participant to feel as if his home is being "invaded" or "overrun with experimenters." If many product team members want to view the home interview, you should find alternate ways to accomplish this. One option is to train development team members to collect data. Another option is to solicit permission to videotape the home interview for development team members to view. A third option is to conduct a follow-up focus group session with several participants at the lab and permit development team members to observe or participate.

We suggest that you have several experimenters or teams trained to conduct the in-home studies. The testing hours, typically 5:00 to 8:30 p.m., are somewhat inconvenient for experimenters who work normal business hours. Sharing the testing across several experimenters will reduce the burden of these hours and will also help ensure that all available test times are covered. After all, we do have personal lives, don't we?

Step 4: Putting together a "help desk" or "study hot line"

Participants may need access to telephone support during their period of home use. Telephone support is a tremendous time-and money-saver when compared to making house visits. Additionally, it provides an excellent source of information to you during the period of home use. Following are our suggestions for implementing an effective "help desk":

- Make sure the person who answers the calls is knowledgeable and able to handle any concerns your participants may have.

- Record and tally the calls and issues that are raised. Some of these will identify specific product improvements.

- Set specific and reasonable hours of operation (e.g., 8:00 a.m.–8:00 p.m.), make participants aware of them, and have someone on the other end of the line during these hours.

- Consider a cellular phone and/or pager to permit mobility to the support person (i.e., the support person does not need to wait for calls in a specific location), and it also permits easy sharing across support people.

Step 5: Recruiting participants

As with any product testing, your participants should be selected based on how well they fit the target user profile. If you are working on a unique new product that has either a vaguely defined user or a potential for a number of different user types, generate one or more profiles that you think likely and gain agreement from the development and marketing teams.

A selection criterion that is particularly important to in-home testing is participant home location. You should select only those participants whose homes are within practical driving distance. Also, select groups of participants with homes that are relatively close together, because this will permit multiple home visits in a single day.

For in-home testing, we encourage you to consider using carefully screened company employees as participants. This is potentially more important for in-home testing than for laboratory testing or focus groups for several reasons. First, visiting the homes of participants who are company employees may be safer than visiting the homes of participants recruited by other means. Second, the return of equipment may be more reliable. Finally, there may be more flexibility around scheduling if company employees are able to use company time for your test (i.e., you may be able to test some participants during normal business hours).

To determine whether company employees are appropriate for your test, first consider if you or your client's company is large and diverse enough. If the company is the development team, you will have to recruit elsewhere. If the company consists of 40 engineers, 40 industrial designers, or 40 ergonomists, you will have to recruit elsewhere. If the company has more employees and greater diversity, however, you may be able to find a reasonable participant pool. Following are our suggested minimum criteria for determining if company employees would be appropriate:

- The company has more than 1,000 employees
- The company develops different product lines
- The company sells to more than one type of business

The larger and more diverse, the better, but if the company you are considering meets or exceeds these criteria, you should be able to recruit a reasonable group of participants. When testing film or cameras, for example, we have been able to recruit company employees with photographic experience ranging from a professional photographer to a complete novice (i.e., does not own or use a camera).

Step 6: Deciding where to test your participants

Because we are discussing in-home testing, an easy answer is "in the home." The reality is that you must minimize the time spent in the home, which means you will need to collect as much data as you can at other times. Other opportunities for data-collection include during participant screening, during a phone interview, when the participant is picking up the product, and when the product is first delivered to the home.

Step 7: Determine the best way to get the product to the home

When considering how to get the product under evaluation to the home, you should assess several factors, including size and portability, installation complexity, and the importance of user installation in your test. Several scenarios are described below along with their benefits.

Participant picks up the product and takes it home. In this scenario the participant comes to you to pick up the product. This is the best overall scenario, as it saves you, the experimenter, time in not having to travel to the home at the start. Additionally, it provides an excellent opportunity to collect data. One can also conduct unpack/setup testing at this time if it is important to observe behaviors related to the setup of the device. The advantage is that your participants are more likely to be fully enabled when they get the products home. One potential drawback, however, is adequately simulating a realistic home environment for setup.

Product is delivered to the home. This scenario may be best used with large or fragile products. Large products that would be delivered to the home under normal sales circumstances should obviously be delivered by you for your evaluation. Fragile items including computer equipment, engineering models, and other systems that may not even have proper shipping materials available are also best delivered to the home. The time of delivery provides an opportunity to collect early interview data and home environment data, which can reduce the time required for conducting the final in-home interview. One can also observe the participant setting up and even first using the product.

Product is delivered to the home and installed for participant. If unpack and setup is not a priority, and the most important aspect of your test is to have the participant using the product quickly, a time should be scheduled to deliver and set up the product. Delivery and installation provides most of the opportunities listed in "Product is delivered to the home" above.

Step 8: Initial interviews, early data-collection, and training

Initial interviews with participants are required to explain the study objectives and expectations. These interviews are also an ideal time to gather participant demographics, product expectations, and early product use data. Some of this information must be collected prior to the home use anyway (e.g., product expectations), and some of it should be collected to save time during the actual in-home test (e.g., demographics). These interviews can be performed either before or during product pickup, delivery, or installation. Following is a description of the various types of information that may be gathered.

Information that must be captured prior to home use of your product.

Product expectations. This information can be captured only prior to product exposure and use. It includes a broad range of items such as how do participants envision the product being used, what quality do they expect, what speed do they expect, how much fun do they expect it to be, and so on. This information should be collected in a way that permits comparison to similar questions, administered after use, that probe how well the product exceeded, met, or fell short of expectations.

Current use behaviors. If the product under evaluation replaces a current product or provides a more convenient solution, it may be important to understand current use patterns of the participant. This data can be captured through simple questions, by describing use situations and eliciting responses, or by soliciting use situations directly from the participant.

Information that prepares the participant for the period of home use

Study objectives. Study objectives should be clearly explained to each participant prior to the use period. Because testing does not occur in the lab and the experimenter is absent for most of the use period, it is important to describe the objectives during the initial interview and when the product is picked up or installed. A written set of study objectives should also be provided for participants to reference at home during the use period.

Product demonstration. Although it may appear to undermine an objective of having participants learn and use a product on their own, a demonstration of the product may be practical if not representative of what is expected in a sales environment. It is often important that the participant at least be aware of the product's capabilities, which may not be a biasing factor. The demonstration can mimic a 30-second commercial in which a message about the product and its purpose is conveyed, or it can mimic the type of demonstration a customer would likely receive at a retail store. To ensure that the demonstration does not show the participant how to use the product, you can either describe features, show examples of results, or demonstrate features while concealing the steps involved.

Notebook/diary use training. Many user issues may occur in the first days of use when the participant is installing and first using the product, and other issues may occur anytime during the home use period. A notebook/diary provides a place for the participant to record these issues as they occur and before they are forgotten.

Provide a notebook/diary to all participants and show them how to use it. Include examples of content and format, and explain the importance of recording these issues as they occur.

With or without training, the information we have received from a notebook/diary has been quite valuable. With or without training, it is also quite variable, ranging from nothing, to short "bullet lists," to pages of notes which include impressions and expectations. Despite its variability, it often successfully documents issues that occur during the use period that might not otherwise have been captured.

Because of the variability and occasional difficulty in interpretation, this notebook should always be reviewed with the participant during the in-home interview that occurs at the end of the use period. Because of the time that has elapsed between first use and the in-home interview, revisiting the issues recorded in the notebook brings issues back to mind. This, in turn, helps participants provide more complete impressions of the product during the final in-home interview.

Unpack/setup test. If the setup is expected to be challenging, have the participants set up the product in a simulated environment in the lab, or have them set up the product when you deliver it to their home. This will save phone calls and ensure that your users are enabled so they use your product as much as possible.

If the user is picking up the product, this test can be scheduled in the lab. If the product is being delivered, this test can be scheduled at the home when delivery occurs.

Encourage using the help desk. Provide your participants with a telephone contact and encourage them to use it if they experience a problem with the product or need additional information.

Put it all in a single package. Organize all materials that a participant will need in a single package or kit. Include the study objectives, user manuals, logbook, help-line phone numbers, and other accessories. This gives your participant "one place to look," and it helps you when packing materials at the conclusion of the in-home study.

Save the packaging! It is also wise to tell the participants to save all the packaging and materials you supplied them with. This will help ensure safe transport of the product from the home to the lab. Also, products under evaluation often go to several homes sequentially, and reusing packaging materials can save as much time as it does trees.

Schedule the home visit. The home visit should be scheduled either during the initial interview or when the product is picked up or installed. This reinforces the use period to the participant and avoids repeated phone calls, during the use period, to find a mutually agreeable time for the home visit.

Information you need to capture before the interview is concluded

Directions. Use the initial interview to get clear and legible directions to the home. Sometimes participants are able to sketch a map or give you directions on a local map, which will be of value when you travel to unfamiliar parts of town.

Adequately document equipment that goes to each home (e.g., record serial numbers). This will assist in tracking equipment, and you will also have proper documentation if equipment is stolen from a participant's home. Additionally, this documentation permits you to track and review the performance of specific units if this becomes important over the course of testing.

Other information that is beneficial to capture during the interview

Additional demographic information. There are typically two levels of demographic information collected. The first level satisfies the screening criteria for participant participation. The second level provides more information that might be of interest or could assist in explaining the data that is collected. This demographic information could range from family size to salary range to other product use. Although it is not necessary to collect this information during the initial interview, it is better to capture it early rather than during the final in-home visit when you are attempting to minimize time at each home.

Step 9: Period of home use

This can be the most insecure time for you, the experimenter. Although you would like to be at the home when the participant is using the product, you cannot. Although you would like the participant to turn on a video camera every time he uses the product, he won't. Still, you know that data-collection is "ongoing" during this period of home use. Whether or not you use software and modems to record and monitor use, you will find the following methods to be both efficient and practical.

Telephone interviews. A call to a home several days after product installation can determine if participants are experiencing any problems that are preventing them from using the product. At times, it even makes sense to schedule a time to conduct a telephone interview with the participant at home in front of the product. Your objectives for calling may include the following:

~ Understand and solve problems

~ Ensure that the user is hooked up and using the product

~ Reinforce study objectives and expectations of the participant

~ Encourage use of the notebook/diary

~ Determine which features have and have not been used.

The format of your phone call should be relaxed, permitting the participant to discuss what is important to them.

Telephone support calls. As mentioned in an earlier section, these calls can yield useful data. Participants usually call when they are experiencing problems, and your job is to understand and help them solve the problems. This data is usually quite obvious, but tallying the problems and reviewing them at the end of the study may identify specific product improvements.

Interim home visit. To ensure the product is fully and properly set up, an early home visit may be necessary. These visits can be reduced or avoided by scheduling and conducting a telephone interview or by having participants unpack and set up the product in the laboratory. But with some complex situations, an interim visit may be the only way to ensure that the product under evaluation is properly and fully set up.

Service visit. At times, a problem cannot be solved over the phone, and one of the experimenters will need to visit the participant's home. It could be a training issue that cannot be handled over the phone, or equipment that needs to be tested and repaired or exchanged.

Step 10: The final home visit/interview

In preparation of the home visit, it is important to:

Make a reminder call to each participant one day before the home visit. During this call, reinforce your arrival time, how long you plan to be in their home, what they should prepare, and that you will be removing the equipment at the end of your visit. Additionally, clearly instruct the participant to leave the product in the place where they have been using it because you plan to use it during the visit. Otherwise, a participant anticipating removal of the equipment is likely to pack up the product before your arrival.

Make a phone call prior to your arrival at the participant's home. This will also help ensure the participant is expecting you and ready to begin the interview.

Travel together. When traveling to the participant's home, interview teams should travel together. This will ensure same-time arrival, and it is better to be lost together than to be lost separately.

Clearly identify yourself. When you arrive at the participant's home, wear and show a picture identification and state your name to the person who answers the door. This is important because the interview team may not have had prior contact with the participant, and the person answering the door may not be the participant.

Do not overstay your welcome. Plan to complete your study and be out of the participant's home in one hour or less. Remember that participants tend to become fatigued after an hour of testing. Whatever test duration is established, however, it is important not to exceed it. The test duration, to the participant, is the time you spend in the participant's home. You must therefore consider introductions, equipment start-up, and equipment packup and removal, along with the administration of study questions.

When you enter the participant's home, you will likely bring other equipment with you for demonstration and data-collection. Be courteous and informative. Show and explain to the participant any equipment you will be using (light meters, data recording equipment, etc.).

Politely decline food and drink. You are in the home as an interviewer, not a guest. It is important to reinforce this; time the participant spends filling entertainment requests is lost time. If a participant is insistent, accept the offer rather than offend them. Some cultures may find your refusal offensive if you do not accept their offer.

When you begin your evaluations, review the study objectives and what you plan to accomplish during this visit. Reinforce that negative and positive comments regarding the product are of equal value. They are helping you to understand how the product will be used, and to figure out how to design the product to be the "best it can be." Always leave the option to change the design open to them.

Ask the participants to set up the product and room, and position him or herself for normal use. This will establish the normal use environment, prepare for demonstration of use by the participant, and permit collection of home environment data (i.e., viewing distance, lighting conditions, etc.).

Split up the work and perform it concurrently. It is important that each experimenter's tasks are well defined so that information is not missed. If you have two experimenters, split the tasks so that one experimenter serves as the primary interviewer and the other experimenter handles other tasks (e.g., collect home environment data, document the product positioning, review the notebook/diary, and pack and remove the product). This separation of tasks tends to keep both experimenters busy, and it also permits tasks to be completed concurrently, thus minimizing the duration of the home visit.

Following are considerations regarding the data you are likely to collect:

Home environment data. Document everything that is important about the home environment. This could include measuring room size, lighting, participant-to-product distance, cabling and hookup, and integration with other products. Taking pictures and video is another way to document the home environment. It is very important, however, to determine if the participants are comfortable with your taking pictures or video. If they are comfortable with it, you should have a prepared consent and model release form available for them to sign.

If the participant is not comfortable with pictures or video, data can also be collected by simply drawing pictures that describe aspects of the environment that are important to your test.

Notebook/diary. Review the notebook with the participant. Because this notebook can provide an excellent accounting of learning issues, usability issues, and problems that occurred over extended use, it is important to clarify and fully understand what was written.

Additionally, this review also raises issues that may have been forgotten by the participant for discussion when they are answering other interview questions.

Use behaviors. Review use behaviors reported during early interviews and discuss how the new product changes them. Discuss use behaviors in two distinct parts: (1) what occurred during the period of home use, and (2) what they anticipate in the long run if they owned the product.

Product expectations. Ask questions to understand how well the product exceeded, met, or fell short of expectations on the variables that are important to your study. Use the same scales that were used during the early interview when probing product expectations so that you can make comparisons. Also, ask whether aspects of the product exceeded, met, or fell short of expectations. Such questions tend to elicit very specific and valuable comments about the product.

Purchase interest and price sensitivity. Ask the participants about their interest in, and intent to purchase, the product they tested. Because they have had a period of home use and have attempted to fit the product into their lives, they are likely to give you thoughtful answers and a number of useful comments.

Ask participants to demonstrate use. It is often useful to run the participant through a usability protocol during the home visit, asking them to perform certain operations with the product. Because there may not be time to evaluate all features, you may need to select specific functions that are in question or are considered most important. It may be more useful to run a looser usability protocol in which you first probe the participants on how they used the product, and then ask them to perform a "scenario" that they just defined for you.

Handling others wanting to make comments. Family members or friends may be present during the in-home visit. Decide up front where to focus the data-collection. If you plan to collect data from everyone present during the interview, you will need to design your data sheets to accommodate this, distinguishing between the primary participant and others. If you plan to collect data from only one primary participant, you will need to explain your primary focus. Either way you collect the data, however, everyone present will need to feel welcome and involved.

Packing it up. At the completion of your evaluation, pack up the product and accessories using the same packing materials used to transport the product to the home. Take inventory to avoid leaving anything behind. Your test product is ready for the next home.

Concluding remarks

In this chapter we have discussed the criteria for determining if an "in-home" test is feasible, and we have described a ten-step process for successfully planning and executing an in-home study. While our methods may not apply to every situation, we feel it provides practical guidelines, evolved from a number of experiences that will assist in many situations. With careful planning, in-home usability testing can be an extremely valuable method to gather information from your users. We wish you much success with your testing; feel free to contact the authors with your experiences.

Remote Usability Evaluation Over the Internet

Ron Perkins,
Design Perspectives

Introduction

One of my clients wanted to quickly understand how people might use their Web site before taking it live. They also wanted a geographically diverse population to test it. I had heard of remote testing at Sun Microsystems (Hammontree, M., Weiler, P., & Nayak, N., 1994) and decided to work out a way to do the same thing, this time using the Internet. With the explosion of the WWW and the diverse content and applications created every day, it seems appropriate to explore how the medium itself can be exploited to better understand design issues.

It is possible to watch people use software applications and Web sites without actually meeting the test participants by using the Internet for remote evaluations. This paper focuses on live remote testing as one means of evaluation. Credit for the work presented here on the taxonomy of remote evaluations should go to the workshop attendees on Remote Testing at UPA '97.

Much work has been done with remote testing at Sun, IBM, and Virginia Tech (Hartson, et al, 1996), but most of it has been done on internal companywide networks. This article presents an overview of remote evaluations, discusses some tools and techniques, and compares benefits and challenges to lab-based testing with a focus on live remote testing. I would like to emphasize that this is not about how remote testing should replace lab testing; each method has advantages and disadvantages.

Live remote testing can be an effective way to gather usability data quickly, conveniently, and inexpensively by using the Internet as a viewing tool. It enables you to expand the demographics of your participant base to any part of the world. Testing either Web sites or software applications can be done simply by arranging a phone call with a participant, having them download the test application or view the site, and then interviewing them over the phone while watching the participant's computer screen with special software. You can even get good behavioral data without seeing the participant's actual screen in a low-cost method, described below.

Thus, you get behavioral data in real time with the opportunity to talk directly with the participant while they are in their own environment. With existing tools and technology you can see exactly what the user is doing on their screen, in their office, and on their platform /resolution/color palette.

Remote usability evaluation methods

Several methods are currently used to perform remote evaluations for specific purposes with varying success. Four distinct types can be classified as:

- Live or collaborative remote testing

- Surveys

- Automated usage tracking

- User-reported critical incidents.

These types were reported by Castillo (1997) and identified at the UPA97 workshop. A brief description of each method and when it would be considered useful is included in a matrix at the end of this article. Of the four types, I will focus on "live remote testing" and argue that, while it currently presents some challenges, the benefits are worth the effort. The live remote testing method offers a means to get qualitative information similar to laboratory tests but with greater speed and access to more diverse participants. It is one way to supplement more quantitative data than you may get from the other remote evaluation methods.

Live remote testing: how it works

After recruiting and arranging for an interview time, you call the participant and interview them over the phone while watching their computer screen with special software over the Internet (viewers described below). Although it is possible to talk over the Internet, the sound quality is far better using the telephone. Similar to a laboratory usability test, you interview the subject, collect rating data, and watch them perform tasks. By videotaping the screen on which you are following the user, you have a record of the session.

Viewing Tools

Screen viewers. You can watch the user in real time, even as he/she moves the cursor with a fast Internet connection and software that allows connecting to the user's IP address. The user downloads the software (some of it is freeware), installs it, and reads their IP address to you over the phone. Some examples are Look@Me, Timbuktu (both from Farallon), and Copycat (IBM). This is what you would see of the user's screen—the desktop is shown in the lower window and can be made fullscreen:

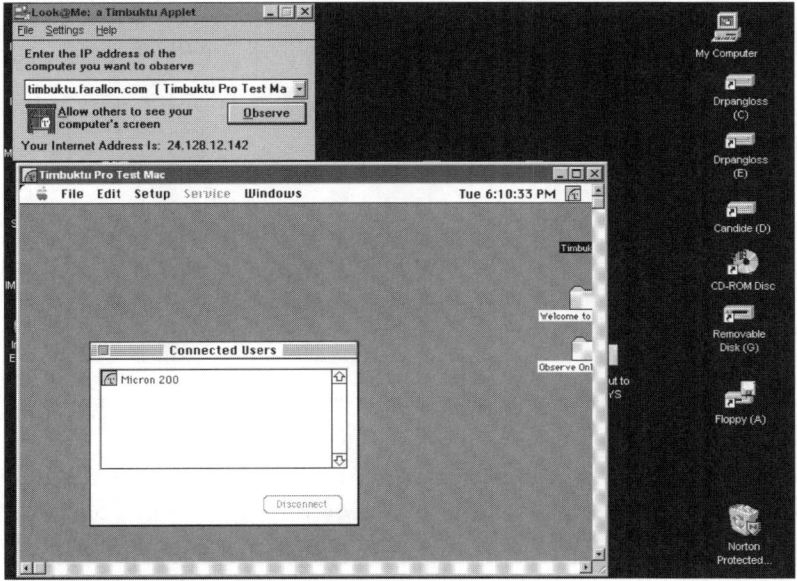

Figure 1. Win95 screen shot of Look@Me, viewing a remote Macintosh desktop.

Viewers are unobtrusive and simple to use but there are problems. One major drawback of the screen viewers is that they require no firewall at the participant end. I have found it difficult to convince network administrators to allow open ports, even temporary ones. Broadband communications like MediaOne *Express%o* now offer high speed without the firewall restrictions. Another viewer that I have used, albeit with some reliability problems, is Lotus SameTime™. It works without restrictions from a firewall and can be used for free at http://stdemo.lotus.com.

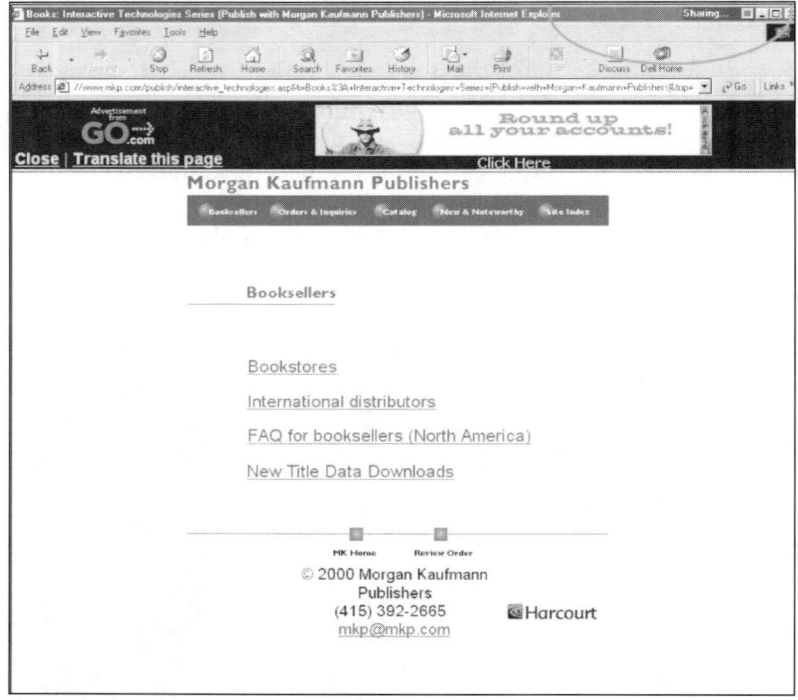

Figure 2. Sharing a Web browser window with Lotus SameTime (the word "Sharing" appears in the upper-right corner to indicate that the application is being remotely viewed).

Conferencing software. Another way to interact with the user is through videoconferencing software such as NetMeeting (Microsoft), ShowMe (Sun Microsystems), CUSeeMe (White Pine), or Proshare (Intel). With enough network bandwidth, you may even be able to see a scratchy picture of the person while they are participating in the test. While firewalls are not usually a problem here, the setup and ease of use of the conferencing software can become a factor in running tests. Also, the participant has to own and install the licensed software. There are a few

Web services that offer similar functionality to NetMeeting without the requirement to install clients. One is at www.webex.com and can be used for online conferencing.

Vividence provides a complementary service to "regular" usability evaluation. It may work well for an initial assessment of a site for redesign. They have a proprietary browser that has some special buttons such as "Answer" and "Comment," allowing them to collect more specific data than a commercial browser. The data that is collected may allow you to see what appears to be a problem but not *why* it is a problem. Vividence can be found at http://www.vividence.com.

Some contrasts with conventional lab testing

There are significant differences between remote testing and conventional lab testing. Let me repeat—this is not an argument for preferring remote testing over lab testing. The data you collect may be similar but with some important differences. Some of the benefits over lab testing are lower cost, participant convenience, and the ability to reach geographically diverse users. Some new challenges arise that are both technical and methodological.

Benefits

- **Natural environment.** The user is in their natural environment rather than in an artificial setting such as a laboratory with people watching them behind a mirror. They are more comfortable, candid, and will behave more realistically.

- **Technical environment.** A participant may have a very different setup for your product in their office or home, one you did not anticipate. You will see their monitor color and resolution settings, browser, and connection speeds.

- **Unrestricted demographics.** Since there are no geographical restrictions, you can reach more and diverse users without traveling to meet them. You may need a translator for other countries.

- **Time savings.** A test can be run both quicker and with less cost. Recruiting is easier because participants do not have to agree to come to your lab. Although the basic test administration tasks, such as study design and stimuli setup, are the same, less time is spent adjusting lab equipment, resetting computers after each subject, and dealing with social amenities to make the subject comfortable.

- **Cost savings.** You can compensate subjects less to participate as they are interviewed in their office (probably being paid by someone else) and do not have to travel.

- **Easy recruiting.** Recruiting is the single hardest thing to do whether for lab testing or remote tests. An argument could be made for remote testing that recruiting by e-mail is more efficient than telephoning. Of course, if you are not testing Web sites, you have a skewed population of users that have e-mail.

- **Facility cost savings.** Another cost reducer is that you do not need any lab space, mirrors, scan converters, and other equipment like movable cameras. You can record both the voice and screen with a camcorder at your location. You can see the person on the computer screen if you use conferencing software. Overall, it costs less to set up because of its simplicity. With broadband, testing can be done from a home office. Remote observers can be included by conferencing them in on the telephone and having each remote observer site also watch the participant's computer (via screen viewers mentioned above). This reduces travel costs and allows more observers to get involved.

- **Low-tech/low-cost method.** For low-bandwidth modem users, remote testing can be done by simply talking on the phone while following the user on your computer screen without a screen viewer. Although you do not see their screen, you may detect confusion by their words and tone. This method forces them to describe exactly what they are doing in real time, effectively thinking out loud. The challenges are to keep in sync with the user; make sure you adjust your computer for their screen size so that you are both looking at the same things.

Challenges

There are some significant questions and drawbacks to remote usability evaluation, both in general and over the Internet:

- **Bandwidth.** Currently, although screen viewers work over a modem, they degrade the speed both for the test participant and observers. You need an ISDN, T1, or better Internet connection to see what the user is doing in real time.

~ **Firewalls.** It is increasingly rare to find anyone with a T1 and no firewall. One of the greatest challenges to live remote evaluations is to find a way to convince network administrators to let the user out from behind the firewall temporarily.

~ **Who is the user?** Without conferencing software, you will not see the user while testing them. Missing data in the form of facial expressions may be a problem for some tests.

~ **Unwanted participants.** It may not be clear who is with the participant on the other end; you could be inadvertently testing one user in front of an auditorium full of your competitor's designers!

~ **International issues.** There are language and recruiting problems when testing internationally. It is challenging enough to do a remote test, even more so through a translator.

~ **New technical logistics.** The difficult technical hurdles you may encounter are successfully getting the participant to download and install screen viewers, or even worse—install and manage videoconferencing software.

~ **Low-tech/low-cost method:** staying in sync with user. When you cannot see even the user's screen, you are dependent on him/her to tell you about mistakes or where s/he is going while you talk with them on the phone.

More resources

There is an emerging body of information on remote usability evaluation. The Remote Evaluation workshop will be repeated this year at UPA '98 in Washington, D.C.

The workshop on Remote Testing over the Internet held at UPA '97 helped to identify a taxonomy of methods. Workshop members included specialists from US West, Microsoft, IBM, AT&T, Sun, and a number of other practitioner/researchers from smaller companies who were doing remote testing. The workshop highlighted a list of available tools and existing, classified remote evaluation methods. These results are captured in a Web page which includes notes from the workshop, tools, a bibliography, and contact names for further information.

Remote usability evaluation Web page

Interested parties are invited to contribute their experiences on an emerging Web site put together by workshop member José C. Castillo, available at:

http://miso.cs.vt.edu/~usab/remote/

Other remote evaluation methods:

- Automated usage tracking is a quantitative tracking of user events such as mouse clicks and keystrokes.

- Electronic surveys are questions that are built into the application being evaluated; results are sent back to developers over the Internet.

- User-reported critical incidents are the collection of usability data centered on critical incidents, self-reported by the user, with special software embedded in the application being evaluated.

Method	Description	Technical Requirements	Data Collection	Analysis	Limits	Tools
Live Remote Testing	Real time observation one-on-one interviews or co-discovery method. One to many observers of behavior over the Internet	• Audio (phone/Internet) • Video (optional) • Connection - Internet, dedicated dial-up; recording device (audio/visual); special software	• High speed connection - same as lab • Low speed connection - no person video, may follow user on screen. • Real time observation of users	Same as lab and field testing	• Technology (not there yet) • User has to install viewer • Real time scheduling • With low technology - lose nonverbal behavior	CopyCat, Look@Me, LapLink, NetMeeting, CUSeeMe, ProShare
Automated Usage Tracking	Quantitative tracking of user events	• Special logging software or special version of application • Data collection software • Data analysis software	• Time shifted collection • High volume • What, not Why - lower quality data • Quantitative	• Special software • Special skills • Need controls	• Lose context • Lack of control • Resource intensive • Not enough on its own	• WinWhatWhere • Web Server Logs
Surveys	Embed questions in tool and get results over Internet	• Get survey to user • Get it back	• Preference / subjective data • High volume, quantitative • Return rate dependent	• Lower technology • Can use higher technology	• Not enough on its own • Response dependent • Could be self-selected	• SUMI, Quis, PSSUS, Group Systems (Ventana) • Could write your own
User-Reported Critical Incidents	Selective data-collection initiated by the user	Special software included with software application	Critical incidents identified by user during task performance	Less analysis required	• Self-reported • Response dependent • Special Software	• IBM Screen Cam • Create your own

Table 1. Four methods for remote usability evaluation.

References

Abelow, D. (1993). "Automating Feedback on Software Product Use." CASE Trends (December), pp. 15–17.

Castillo, J.C. (1997). "The User-Reported Critical Incident Method for Remote Usability Evaluation." MS Thesis, Department of Computer Science, Virginia Tech, Blacksburg, VA 24061, USA.

Elgin, Bruce (1995). "Subjective Usability Feedback from the Field over a Network." *SIGCHI Bulletin* (October 27 [4]), pp. 43–44.

Hammontree, M., Weiler, P., & Nayak, N. (1994). Remote Usability Testing. *Interactions* (July), pp. 21–25.

Hartson, H.R., Castillo, J.C., Kelso, J., Kamler, J., & Neale, W.C. (1996). "Remote Evaluation: The Network as an Extension of the Usability Laboratory." Proceedings of CHI '96 Human Factors in Computing Systems, pp. 228–235.

Siochi, A.C. & Ehrich, R.W. (1991). "Computer Analysis of User Interfaces Based on Repetition in Transcripts of User Sessions." *ACM Transactions on Information Systems,* (October, 9 [4]), pp. 309–335.

This link contains a table of remote tools compiled by Alice Preston: <http://www.stc.org/pics/usability/newsletter/9901-remote-tools.html>.

Alice Preston, "Usability Engineer, Bellcore." Reprinted from *Usability Interface,* Vol. 5, No. 3, January 1999.

Tools and sites for automated usage logs and critical incident reporting:

ErgoLight, where you bundle the recorder with your software, user does testing, then sends log back to you for analysis. See www.ergolight-sw.com.

The author does not endorse or personally recommend any of the above tools/sites—use your judgment regarding the validity of the method/data that you collect.

How (Much) to Intervene in a Usability Testing Session

Howard Tamler,
HT Consulting

During the summer of 1997 on the UTEST forum (a popular listserve for usability professionals), a Stanford University researcher unequivocally weighed in against any intervention by the tester in a usability test session. Invoking the standard procedures for experimental research in behavioral science, this individual argued that passive, unobtrusive observation by the tester was necessary to provide a level of experimental control which ensures a uniform experience for all test users, as well to avoid experimenter bias (the conscious or unconscious influencing of the user's behavior by the expectations of the tester).

Though this problematic issue has come up before in different guises, it again generated a fair amount of discussion, all of which bears on the wider question of how much (and what kind of) intervention is useful during usability testing sessions. This issue has not only been debated on UTEST but has also come up in tactical discussions with several of my clients; I believe it needs to be addressed in any usability test, at least tacitly if not deliberately.

Nielsen (1993) is pragmatically flexible on this issue, but basically counsels reticence. His initial advice is to avoid interacting with the user, making only "uncommitted sounds like 'uh-huh' to acknowledge comments from the user and to keep the user going. . . ." He advises departing from this rule in only two situations. One situation is when users are frustratingly stuck; in this case, helping them out is useful for getting the test moving again, reducing their stress so that they are able to proceed rationally, and restoring their self-esteem. The other situation is when they are grappling with a familiar problem with a well-known impact based on observations of previous test users, in which case there is little point in sitting through the same problem repeatedly. In contrast, in a later section on "thinking aloud" studies, Nielsen advocates prompting the user as needed with questions such as, "What are you thinking about?" "What do you think this message means?" or "Is that what you expected would happen?" However, these interventions should be reactive rather than proactive, that is, limited to situations where the user has already noticed or commented on something.

A quite different view was developed in an important foundational paper by Whiteside, et al, (1988), who argue that what can be measured or objectively observed doesn't begin to do justice to the user's underlying experience of usability or difficulty, and, therefore, operational definitions of usability are necessarily reductive and limited. In other words, merely observing a user's behavior is not enough to understand what's happening. As they put it, "If usability engineers . . . base their interpretation on those observations, they will report their own experience of the user's behavior, not the user's experience," and ". . . cannot know if conclusions are warranted unless he or she shares these interpretations with the user."

Hence, the authors advocate treating a user as a subject and full partner in the study rather than merely as an object to be observed. In short, the user's subjective experience is the ultimate (and most comprehensive) criterion of usability, but the only way to access this experience is to talk with the user. While the authors make these points to justify "contextual inquiry" as an alternative to controlled laboratory testing, I believe that their ideas apply with equal justice to usability testing.

All things considered, I suggest that a useful way to look at this issue is in terms of the purpose of a usability test. If the purpose is scientific hypothesis testing or collecting quantitative data for comparison against other products or usability goals, then experimental control, including non-intervention, is of paramount importance.

For example, the process of responding to a tester's questions would clearly invalidate various quantitative measurements, such as task completion time. However, if the purpose is to gather qualitative data that will enable us to identify significant user interface problems and recommend design solutions, then openly interacting with users in various ways may not only be useful, but also sometimes necessary for collecting and understanding the required diagnostic information. In short, because scientific research and formative evaluation may have significantly different goals, quite different standards can apply.

With diagnostic usability testing in mind, the following kinds of interventions can be useful during a test session:

Probing questions

First, if there are particular questions or areas of the user interface that need to be addressed, but the user does not spontaneously interact with these, it may be necessary to focus the user's attention on these features by asking pointed questions, such as, "What do you think those buttons are for?" or "How can you make that window fill up the screen?" or "If you needed help, what topics would you look under?"

Second, Spool (1997) suggests that questions are the only effective way to elicit and clarify a user's mental model, giving the example of users who never clicked on an Apply button because they thought it meant "apply voltage," something that Spool found out only by asking them. Likewise, I tested an application for organizing graphics files by grouping them into "albums" consisting of miniature (thumbnail) pictures of each file. When a thumbnail was inserted into an album, it was not transparently clear that only a link between the album and the graphics file was being created, since the drag-and-drop transaction was visually identical to moving or copying the graphics file to the album. Hence, the only way I could uncover the user's model of this transaction was to ask them to explain what they thought was actually happening.

Third, regardless of the content of questions, the phrasing can have a great impact on how tactful and effective the questions are. In this respect, a good rule is to avoid "why" questions, because they are not only imprecise but also imply criticism. For example, if I say to my shivering and crying child, "Why didn't you wear your jacket?" she knows she's being scolded rather than being asked for information. Likewise, "Why did you select that file?" suggests that the user needs to justify her action because it's incorrect, and does not specify what the questioner is fishing for. A better way to phrase the same question is, "When you selected that file, what were you expecting to do with it?" or "How did you decide to select that file rather than some other file?"

These more neutral and precise questions imply a sincere request for particular information, rather than a request for justification. The difference is subtle but can often be potent in terms of how the user responds.

More generally, Dumas and Redish (1993) show how bias can be avoided by asking questions which are neutral and open-ended, rather than leading yes/no questions which unwittingly express your opinion and thereby invite the user to agree. For example, they advise asking, "What are you trying to do?" (open-ended) rather than, "Are you trying to copy that file?" (leading). As another example, they advise asking, "How were those instructions in terms of being either clear or confusing?" (unbiased) rather than, "How confusing were those instructions?" (biased, since only "confusing" is mentioned). A third example is asking, "What other information do you need to make that easier?" (open-ended) rather than, "Would adding information X make it easier?" (leading the user to endorse X). Of course, if you need users' opinions about adding X, you can ask them a leading question, but only after asking them an open-ended question first, to see what they come up with on their own. (For more such examples, see the very useful table on page 299 of Dumas and Redish.)

A related tactic is useful when there is more than one method for accomplishing a task (which is usually the case), and you are interested in seeing how well users can figure out a particular method, but the user finds some shortcut instead. In this case you can ask the user to redo the task but to use a different method.

Active listening

Active listening, apparently overlooked in the usability literature, refers to giving feedback to the speaker which demonstrates caring, nonjudgmental acceptance, and understanding on the part of the listener. What it boils down to is empathy—feeding back the speaker's meaning by paraphrasing it in the listener's own words.

For example, I did a "think aloud" study of a utility program which helped users recover space on their hard disks by identifying unnecessary files, accumulating them in a list, and finally deleting the resulting list. One user continued to remove unwanted files from the accumulator using the Remove button (which was counterproductive, unknown to him), and asked what the Add button was for. To make sure I understood his mental model, I said something like, "In other words, the Add button doesn't make sense to you, since the whole point is to remove files, and obviously the way to get rid of files is to select them and then click the Remove button." The user responded "Yes, that's it," thereby letting me know that my interpretation of his mental model was correct.

Now, if I had just listened and made notes of his comments, which were somewhat confused and repetitive, I might have been misled. Or, if I had asked a clarifying question such as, "What's the purpose of the Remove button?" or "So, what's the basic procedure for getting rid of files?" I would risk irritating the user who had been telling me all along in his own way, and who might start to wonder if I was really paying attention. And if his answer to my questions was as inarticulate as his previous monologue, I might still be left unsure of his meaning.

Instead, feeding back my tentative understanding as an articulate statement demonstrated that I was observing carefully, empathizing with his perceptions, and trying to make sense of them. Likewise, his confirmation reassured me once and for all that my hypothesis was correct. Moreover, carefully paraphrasing a user's tentative or confused comments often helps clarify the problem in the user's mind as well. In this regard, there have been occasions when users told me that my feedback had helped them better identify their problem or had expressed it better than they could.

Hence, in this situation, active listening was an effective and efficient way of getting inside the user's head, which helped to clarify and confirm my understanding of his thinking. This is generally important because users are often inarticulate, and communication is often ambiguous or otherwise difficult even under the best circumstances. In addition, because active listening builds trust and rapport in the user, it also reinforces and encourages thinking aloud, and does this in a way that's more tactful than a blunt command such as, "Please tell me what you're thinking now." For reasons like this, I consider active listening to be the single most-powerful communication technique for achieving the kind of empathy advocated by Whiteside, et al.

Although the examples above may appear simple or even trivial, active listening is not easy to do. For many people, it's a skill that requires considerable effort and practice, especially since it goes against the grain of our natural communication tendencies: to evaluate, agree or disagree, ask questions, pursue our own agendas, etc. Partly for this reason, active listening (which I acquired in my training as a psychotherapist) is an essential communication technique for any helping professional (counselor, teacher, lawyer, supervisor, etc.), and has been widely taught as the centerpiece of parent effectiveness training, as well as in many human-relations training programs for managers and supervisory employees. It is also widely parodied as, "I hear you," or something similar and has probably been satirized in "Dilbert" at one time or another.

Finally, while active listening is a neutral way of fostering and demonstrating empathy, there is sometimes a need to explicitly praise or encourage users, especially if they have been having difficulty. For example, the transition between tasks can be punctuated by comments such as, "This is exactly the kind of thing we need to hear, so keep it up," or "You're providing us with lots of good information about the product." The important thing here is to avoid bias; for example, if this kind of feedback is conditioned on the user's having just said something positive about the product, it may reinforce that particular behavior at the expense of reporting negative reactions. However, if it's timed so as not to be associated with any particular response, it can have the desired effect of making the user feel OK and appreciated regardless of what he does or says.

Guiding the user

Usability professionals often advocate letting users figure out a task without any interference, and for good reason. This enables us to see what might happen in the real world, such as how far astray users might go, and how (and whether) they are able to recover on their own from dysfunctional choices. In many cases, this is important information. However, the downside is that sometimes users will never discover the right moves, the task of interest will not get completed or even begun, and the questions which originally motivated that task's inclusion in the test will go unanswered.

The opposite approach, stopping the user from pursuing dead ends or paths which can't lead to task accomplishment, not only saves time but can also ensure that much more ground is covered, especially the parts of the application that are of interest to the tester. That is, telling a user to abandon a wrong path and redirecting him enables the tester to ensure that the task of interest is completed and enables him or her to see what happens at each step.

For example, in a think-aloud study of a utility package consisting of several quite different utilities or subapplications, we were interested in seeing how well users could execute a particular task in a given utility. However, users often couldn't figure out which utility to use for a given task, and one user selected five different utilities before arriving at the correct one. I allowed this user to continue exploring the first utility chosen, but after five or ten minutes, I realized that if I didn't intervene, he might never get to the assigned task. Given the purpose of the test, the essential observation was that this user could not identify the correct utility in the first place, and there was nothing more to gain by allowing him to flounder in the wrong one. Consequently, I told him that this was the wrong utility, so he promptly quit it and tried another one.

Then, each of the subsequent four times when he selected another incorrect utility, I immediately asked what led him to select that utility (if he didn't volunteer the information), and then told him it was wrong. When he finally selected the right one, I backed off and let him figure out the task. In this way, I was able to observe the task of interest without spending an inordinate amount of time. Several of my clients (who have sat with me in the observation room) have remarked that this level of intervention provided them with much more information than they were accustomed to getting with more passive tactics.

Another technique for helping users who are stuck is to give them a hint, just one little piece of information which might be helpful but not obvious. If this hint is sufficient to get the user on the right track, it's a good bet that this information should be somehow provided in the user interface. In any case, helping a user accomplish a task lets users see how it's supposed to work, which can often help them to see what's wrong, missing, or misleading. For example, once a user was told that a backup drive would not work unless it was turned on prior to booting the computer, he was able to immediately (and correctly) point out that there was nothing in the product or documentation that told him that.

Conversely, if users ask for guidance when we want them to figure things out for themselves, we can gently direct the question back to them, or give a non-directive answer. For example, if asked, "How can I create a table of contents?" we can respond, "Well, from what you've seen of this application so far, where could you look for an answer to that?" or "Just do what looks promising to you—I really need to see if you can figure it out or yourself."

Another important place where guidance should be avoided (or evaded) is determining when a task is completed, since knowing when to stop is an integral part of any task.

If you tell users when they are done, for example, you will miss discovering that the application fails to provide sufficient feedback to confirm that a task has been executed. Consequently, it is good practice to explicitly instruct users to let you know when they think they're done with a task.

Finally, when a task has not been completed and the user is either at a loss or has reached the point of diminishing returns, you can terminate the task tactfully by saying something such as, "At this point, you've provided enough information on the problems with this task, so in the interest of time, let's move on to the next one."

User as designer

Spool cautions against asking "design questions" such as, "How would you order the menu options?" since users don't know enough about the particular application, or design in general, to come up with feasible suggestions, and their answers to such questions are generally naive and unproductive. But while asking how the user might redesign a feature violates our cherished slogan that "users are not designers," user input on design may occasionally be useful, and we should be open to it.

For example, an application for engineers included a dialog box that was so confusing, I had to coach several users to get them through the task. After being told how to do the task, three users proposed more or less detailed redesigns, and one of them insisted on drawing it on paper. His design made so much sense that I used a modified version of it in my report, recommending that the client adopt it. In this case, I did not ask users to suggest a design, but they felt so strongly that they spontaneously offered it.

In conclusion, there is a whole continuum of situations ranging from those where it's essential to intervene, to others where passive observation is indicated. As usual, "it depends" on the objectives of the test, on the user's behavior, and on what you need to know in a given situation. My point is that there are contexts where proactive interaction with test users can be a very effective tool.

References

Dumas, J. S. and Redish, J. C.. *A Practical Guide to Usability Testing.* Norwood, NJ: Ablex Publishing Corporation, 1993.

Nielsen, J. *Usability Engineering.* Chestnut Hill, MA: Academic Press, Inc., 1993.

Spool, J. *UIETips.* May 20, 1997, User Interface Engineering Inc., via e-mail.

Whiteside, J., Bennett, J. L., and Holtzblatt, K. "Usability Engineering: Our Experience and Evolution" in *Handbook of Human-Computer Interaction,* edited by Helander, M. New York: Elsevier Science Publishers, 1988.

How Many Participants in a Usability Test Are Enough?

Joe Dumas, American Institutes for Research

Given the present state of knowledge about usability testing methods, identifying the optimum number of participants for a usability test is one-part science, four-parts art, and five-parts common sense. As I talk with usability-testing colleagues around the country, there is some consensus about a few issues that have to do with the number of participants to use in a test and much diverse opinion about the rest. The consensus I hear is that (1) you can learn a lot about the usability of a product with a few test participants, and (2) you should avoid having to use statistical tests of significance whenever you can. In this article, I want to address the issue of how many participants are enough in some detail. I don't pretend to have the last word on the subject, but by sharing my thoughts, I hope to provide a forum for further discussion.

In this article I am not going to emphasize that the participants you choose for a usability test have to be members of the target market for the product you are testing; I assume that readers of *Common Ground* understand that principle.

As with most issues in usability testing, consideration of how many participants to recruit begins at the beginning with the objectives and goals of a test. Let's examine two common usability testing objectives: (1) uncovering most, if not all, of the serious usability flaws with a product or system, and (2) comparing the usability of one product to another or to an earlier version.

Diagnostic usability tests

A diagnostic test is the most common type of usability test. It is a test in which the goal is to find as many usability problems as possible, but to take steps to ensure that you find all of the most severe problems, however you define severe. (In a future article, I will address the issue of what "severe" problems are.) Because there almost always is pressure to keep costs low, part of the skill in planning a diagnostic test is in meeting your objectives with the fewest possible participants. If there will be only one test of a product, then you will want to err on the side of caution and run more participants than you might if you were conducting iterative tests, especially early in design. But what does "more participants" mean here? More than what number? Let's see what the research says.

A series of research papers by Nielsen and Virzi in the early '90s investigated the question regarding how many participants are enough. In all of these studies, test participants performed tasks with products that have a well-defined scope of application; that is, there are a limited number of tasks users can perform with them. Examples of products used in these studies are a voice mail system and an office calendar system. As participants performed usability test tasks, usability testers listed problems as the participants uncovered them. The authors looked at the number of problems uncovered by each participant as participants were added to the test. Here is what they found:

- In all of these studies, the participants varied greatly in the number of problems they uncovered. The least productive participants uncover only 15 to 20 percent of the usability problems; the most productive uncover 50 to 80 percent of the problems.

- There was a lot of redundancy in the identification of problems. Some problems are uncovered by almost all of the participants. In all cases, there are diminishing returns from adding new participants. For example, the first participant might uncover 30 percent of the problems, the second might add only 15 percent more, the third participant might add only 10 percent more, etc.

- Over several experiments, Virzi (1990, 1992) found that about 80 percent of usability problems were uncovered by about five test participants, with 90 percent uncovered by about ten participants. The relationship in the Virzi studies between the number of participants and the percentage of problems uncovered is quite consistent over several types of applications, all with a limited scope. In these studies, Virzi used a Monte Carlo technique to show the general form of the relationship. In brief, this technique shows what would happen if the test were repeated many times. Figure 1 shows a typical graph from this research showing the relationship between the average number of problems uncovered as participants are added to the test.

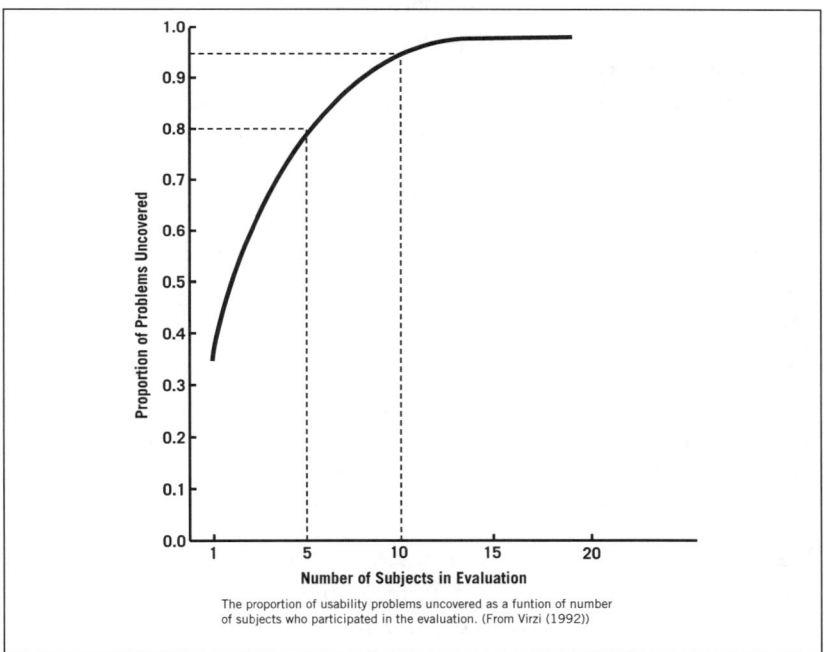

The proportion of usability problems uncovered as a funtion of number of subjects who participated in the evaluation. (From Virzi (1992))

Figure 1. Relationship between number of subjects and problems uncovered.

The curves from this method are quite consistent across several studies in showing that an average test with about five participants would uncover about 80 percent of the problems in a product. When problems are ranked by severity, five participants uncover virtually all of the most severe problems. But with five participants, at least 40 percent of the least severe problems remain undiscovered. Even with ten participants, about 20 percent of the least severe problems remain undiscovered.

The conclusions from these studies seem straightforward:

1. If you run between five and eight participants, most of the time you will find the showstopper problems and some of the others.

2. As you reduce the number of participants below five, you run the risk of missing severe problems, especially if you are unlucky and sample unproductive participants.

3. The cost of uncovering the least severe problems with usability testing is high. Even with ten participants, there will be uncovered problems. This finding suggests that an expert review in which usability experts review the user interface by performing representative tasks is a good supplement to a usability test because the experts are more likely to find the less important problems.

Even though the studies that investigate the number of participants are consistent, there are very few of them. We don't have studies that look at what might be important variables, such as the influence that a particular sample of tasks has on uncovering problems or even whether the order of the tasks makes a difference.

I follow the conclusions from these studies in the planning of my own usability tests. Over the course of 12 years of testing, my experience seems to confirm the research. I almost always feel confident that I have uncovered the most severe problems I would find with five to eight participants in a group.

Ideally, if I had the time and resources, I would test participants until they were not uncovering any more serious problems and then stop testing. But product managers often want to know how long the test will take and that always depends on the number of participants. My rule of thumb is that, when pressed for time, I plan to run five to eight participants in a group for a diagnostic test, nine to twelve participants if there is time. I am confident that for any particular set of tasks, running additional participants will not uncover additional serious problems. But I don't know what would happen if I changed the sample of tasks or the order of the tasks.

There is one study that is not consistent with the others I mentioned. Lewis tried to replicate the findings of these studies with limited success. He did confirm that, as you run more participants, you uncover fewer unique problems. But in his study, five participants uncovered only

55 percent of the problems. In addition, the most severe problems were not uncovered sooner than the less severe ones—a disturbing finding. The product that the participants used in the Lewis study was, however, a commercially available software suite that contained a word processor, spreadsheet, calendar, and an operating system. Participants worked on the suite for six hours and uncovered a total of 145 different usability problems.

There are several conflicting conclusions one can draw from the Lewis (1994) study. One is that the results of the earlier studies cannot be generalized to more complex, more open-ended products. But, one could just as easily argue that the participants in Lewis' study simply had not found all of the problems in the time available. If they had had more time, the data would look similar to the other studies. This is a research question that awaits more study.

I want to reemphasize the point that we have very little in the research literature to justify the common practice of using five or fewer participants in a usability test. The fact that such small samples always are useful at uncovering severe usability problems and that we see the same problems repeated, do not, in themselves, provide evidence that we have found all of the severe problems. Remember that the original Virzi studies (1990, 1992) tested products with a fairly limited scope, such as a voice mail system, and that Lewis (1994) did not replicate Virzi's findings using a product with a much wider scope, an office productivity suite.

Also keep in mind that usability testing is not an efficient evaluation tool for uncovering the less severe problems. In Dumas and Redish (1999), we make a distinction between local and global usability problems, with local problems being limited in scope to one screen or panel. If test participants don't encounter the screen or panel, the test won't uncover the problem. This inability to uncover problems in parts of the product that are not encountered is a weakness of usability testing. We need to continue to supplement usability tests with other usability evaluation methods, such as expert reviews, that are better at finding local problems.

Thus far we have been talking about tests with one group of participants. What happens when a product's users include people with varying experience or demographic characteristics? It is not unusual in tests of computer software that new users of a product include people who are not familiar with an operating system, such as Windows® 95, and people who do have that experience. Do you have to run between five and eight participants in each group? I know of no studies that address this point. In my practice, I run slightly fewer participants in a group as the number of groups increases. If I were planning a test with two groups that differ only in experience, I might start with the assumption that I should run about six in the naïve group and perhaps four in the experienced group. In most cases, the severe problems will be uncovered by almost all of the participants in both groups. In addition, there will be some unique problems that the four participants in the experienced group uncover. The strategy of running participants until you stop uncovering problems also works here.

Comparison usability tests

The number of participants you need for a comparison test depends, as always, on the purpose of the test. If you are interested only in the strengths and weaknesses in usability of each product, the rules that apply to a diagnostic test apply here. With five to eight participants using each product, you can identify 80 percent of all the problems of the products you are comparing.

If your objective, however, is to draw a more quantitative conclusion, determining how many participants is enough is more complex. I certainly would not be confident that I know how much better or worse one product is in comparison with another with only five participants. For example, I would not make a statement that Product A is 20 percent better than Product B based on the data from as few as five to eight participants. I would certainly want to take the variability in the population of users into account. So, for example, I might compute the mean of the average task times for two products and show standard deviations or confidence intervals around those means so that I could see whether the difference between the means is substantial relative to the variability around the mean values. For such a comparison, I would plan to test 15 to 20 participants to feel confident about the stability of my comparisons.

The most troublesome situation in which sample size is at issue is the case where a product developer or a manager of a product developer wants to show that the difference between their product and a competitor's product is "statistically significant." I always try to avoid this situation, but on occasion I have no choice, especially when the client wants to make claims about usability in promotional literature.

Unfortunately, discussions about how many participants are enough to show statistical significance have to include consideration of at least four factors:

- The variability in the population—The more variability there is in the user group, the more difficult it is to find a significant difference. Looked at another way, if you can reduce the variability in the test, you increase the likelihood of finding that one product is significantly better than another. In a comparison test, you can reduce variability by having consistent screening criteria on which to select participants, and having consistent instructions and procedures.

- Setting the level of significance—By convention, the accepted level of significance is .05 (or five chances in 100 that a difference you find could have occurred by chance). The reason that the level of significance is set so high is that, traditionally, scientists have believed that the impact of incorrectly concluding that a difference is significant when it is not is usually worse than not detecting a difference when there is one. If you change the level to .10, you increase the likelihood of finding a difference significant. But you need to have a reason you can justify for changing from the accepted norm of .05. For example, it may be more important in your situation that you avoid concluding there is a difference when there is not one. Consequently, you may feel that it is prudent to change the level to something like .20. If you do make that change, you are more likely to find smaller differences significant.

In a recent article, Wickens (1998) argues a similar point. He recommends increasing the level of significance from .05 in a test of a new product. He argues that when testing is performed early in the design process, it is wise to reduce the likelihood that a good design or design direction will be discarded because the results are not significant. Discarding promising designs is a mistake that eventually hurts users, especially when the designs are evaluated with a small sample size to keep costs low.

- Determining the magnitude of the difference between the products—As the difference in the usability of the products increases, a typical statistical test is more likely to find that difference. All this means is that if one product is clearly more usable than another is, a test of their usability is more likely to be significant. This may sound like circular reasoning because you are trying to measure the difference to reveal its magnitude in the first place. But you do have to make some assumption about the size of the difference you want to detect if you want to compute the size of the sample you need to detect.

- Determining the power of the statistical test you are using—The power of a test is a measure of its ability to detect a difference when there is one. If two electronic mail programs really differ in their usability, the more powerful the statistical test, the more likely it is that the difference will be found, that is, the more likely it is that a statistical test will correctly show that there is a difference.

It turns out that sample size is one of the factors that determines the power of a test. Fortunately, statisticians have worked out the relationship between the power of a test and the sample size you need to achieve it. Two sources that describe the relationship between sample size and power in words that are understandable are Dunlap and Kennedy (1995) and Cohen (1992). They are valuable sources if you need to make a strong case for the sample size you choose for a usability test. But to take advantage of these sources, you will have to make assumptions about the level of significance you choose to set and the size of the difference you want to detect.

For a more extensive discussion of these relationships, there are many books that discuss behavioral science statistics. Two I have found useful are Shavelson's, 1988, Chapter 12, and Kirk's, 1982, Chapter 4.

The good news is that for many of the types of measures we use in a typical usability test, such as average task time, the range of sample sizes needed for significance testing is typically in the 15 to 35 range. My rule of thumb for comparison tests where statistical significance is an issue is not to go below 15 to 20 participants, and to go for 25 to 30 if I can.

Since this article was published, I have conducted three comparison usability tests that came out in favor of one of the tested products. The sample sizes in those studies are consistent with the advice in the original article. In one of these studies, each participant used all three of the products and the sample size was 24 participants. In the other two studies, each participant used only one product. The sample sizes in those two studies were 20 and 36 participants in a group.

There are statistical tests of significance that don't require you to adhere to as many assumptions as the ones we have been discussing. These tests are called "nonparametric" tests (see Shavelson [1988], Chapter 24, or most behavioral science texts on research design). These tests are less well-known, but many of them are simple to compute and they make the issue of statistical significance less mysterious.

I believe that the best strategy in comparison tests, however, is to avoid the issue of statistical significance altogether. If the difference in usability between products is not substantial, using statistical tests of significance is not convincing to anyone. I would rather show that one product is clearly best on each of a variety of measures, such as average and total task time, proportion of completed tasks, ratings of ease of learning or use, and participants' rankings of preference. If all of these measures are not substantially in favor of one product, that product is usually not more usable.

The perception of credibility

Sometimes determining the number of participants in a test is not based on research or statistical reasoning, but on how credible the number of participants sounds to a third party. Marshall McClintock of Microsoft's usability group was the first person I know to use the term "press validity," which refers to testing practices that are followed to answer an anticipated query from a reporter about the validity of a comparison test. For example, you may need only 20 participants to detect a difference of any practical value, but you run 50 participants to influence the press. Or you may include participants from different geographical regions of the country even though these differences are unlikely to have any practical effect.

When it comes to sample size, however, you can run too many participants. As we have seen, a test is more likely to produce statistically significant results the more participants there are in it. The power of most statistical tests increases with sample size. Rather than being impressed when a product difference is significant with 50 participants, I, and I think you, should be skeptical. In that case I am tempted to ask,

Why did it take 50 participants to show this difference? Is the difference really very small; that is, is it statistically significant but of no practical importance?

I hope that you find my rules of thumb useful and that you're stimulated to conduct some research on this topic.

References

Wickens, C., (1998) "Commonsense Statistics." *Ergonomics in Design,* October 1998, pp. 16–22.

Cohen, J. (1992), "A Power Primer." *Psychological Bulletin,* pp. 112, 155–159.

Dunlap, W. P. & Kennedy, R.S. (1995), "Testing for Statistical Power." *Ergonomics in Design,* July, pp. 6–7.

Kirk, R. (1982), *Experimental Design: Procedures for the Behavioral Sciences,* Monterey, CA: Brook/Cole Publishing Company.

Lewis, J. (1994), "Sample Size for Usability Studies: Additional Considerations," *Human Factors,* pp. 36, 368–378.

Nielsen, J. (1990), "Evaluating The Thinking Aloud Technique For Use by Computer Scientists," in Hartson, H. & Hix, D. (eds.). *Advances in Human Computer Interaction: Volume III,* Ablex Publishing Corp.

Shavelson, R. (1988), *Statistical Reasoning for the Behavioral Sciences, 2nd Edition,* Boston, MA: Allyn and Bacon.

Virzi, R. (1990), "Streamlining The Design Process: Running fewer subjects," Proceedings of the Human Factors Society 34th Annual Meeting, pp. 291–294.

Virzi, R. (1992), "Refining the Test Phase of Usability Evaluation: How many subjects is enough?" *Human Factors,* pp. 34, 457–468.

Classifying User Errors in Human-Computer Interactive Tasks

Pawan Vora,
Human Factors Engineer

In interactive usability testing, it is often the errors committed by the users, and not the tasks performed correctly, that help in identifying the source of problems in the user interface.

In my experience, the errors in human-computer interactive tasks are of two types: failure to complete the task and deviations from the ideal path in doing the task.

A common method of dealing with errors in usability testing is simply to count the number of times the users erred. Though straightforward, determining the frequency of errors doesn't provide useful information for improving the interface.

What is required instead is to classify user errors in a manner that pinpoints the source of the problem.

Although considerable research literature on error classifications exists in the area of cognitive psychology, there are no quick-and-easy ways of classifying errors for human-computer interactive tasks that are diagnostic enough to indicate the sources of problems with the user interface. Furthermore, most of these error classifications were designed keeping process industries in mind where typical end users are experts; whereas in usability testing, the target users are typically novices.

In my experience, I have found classifying errors into "errors of omission" and "errors of commission" very useful in identifying the source of problems with interfaces. Note that I have borrowed this from the work of Allan Swain; for more details refer to: Swain, A.D. (1963,

August). A method for performing a human factors reliability analysis. (Monograph SCR-685). Albuquerque, NM: Sandia National Laboratories.

An error of omission occurs when a user omits a step in a task. Examples of omission errors include clicking on the paste text icon without copying the text, quitting the application without saving the work, and so forth.

An error of commission occurs when a user does the task, but does it incorrectly. Types of commission errors are selection errors, sequence errors, qualitative errors, and time errors.

Selection errors, as the name implies, result when the users choose an incorrect option to perform a task. Some examples of selection errors include selecting the wrong menu option, selecting the wrong option in a dialog box, and clicking on the wrong icon (e.g., clicking on the fax icon to print a document).

Sequence errors occur when the user performs the task in an incorrect order. Consider the example of the "merge" feature users have problems understanding specific steps in the procedure or do they need additional information to complete the task successfully.

Another advantage of the proposed scheme is that the investigator can simply look at the user-interaction logs and identify the types of errors that the users made. The categorized error data can be used as an evidence that problems exist and what the problems are. This type of evidence is also easily understood by the clients, project leads, developers, and other team members.

Of course, I agree that this classification is not meant to capture the underlying cognitive processes of the users. But then, is that type of information worth the trade-off in time and effort and is it really useful? Has anyone coded verbal protocols lately?

Reference

Sulain, A.D. (1963, August). "A method for performing a human factors reliability analysis." Monograph SCR-685. Albuquerque, NM: Sandia National Laboratories.

Staking Your "CLAIM" to Usable Corporate UI Design Guidelines

Thomas DiPersio,
US WEST Information
Technologies

Over the last five years, there has been much discussion regarding the creation and use of graphical user interface design (GUI) guidelines as a tool for increasing the usability of application software. Software vendors such as Apple, IBM, Microsoft, and others have published UI design guidelines in support of their own application programming interface (API) architectures. While these guidelines have primarily served the interests of these vendor companies' individual platforms, they were developed, in large part, drawing on research and development results from the field of human-computer interface design (HCI).

Following the lead of software vendors, corporations that develop their own software have adopted UI design guidelines in hopes of increasing the consistency and reducing the development and maintenance time of their application software. The prevailing notion, which is in accordance with accepted HCI principles, has been that less random variance in interface design contributes to more cost-effective development (through more design reuse and fewer design "disagreements") and more usable interfaces. Although formal experimental data supporting these claims are difficult to find, benefits of applying design guidelines are still widely assumed. While usability professionals agree that guidelines alone are not sufficient to ensure usable interface designs, guidelines are foreseen by many, nonetheless, as one part of a suite of usability engineering tools. Consequently, many companies have either adopted published guidelines, developed their own, or, more commonly, established some combination of both.

Unfortunately, the process of creating corporate design guidelines is fraught with pitfalls that can render them little more than a "nice idea" in a neatly bound notebook on the shelf which you point to but do not actually use. Developing usable corporate UI design guidelines is a very difficult task. The main reasons for the difficulty lie in the expressive power and scope of guidelines themselves, as well as the expectations for their use. Guidelines often fall short due to a number of factors that we will explore. Although they are difficult to develop, applying a small number of principles (meta guidelines) during their creation can yield corporate UI design guidelines that are used successfully.

Making sense of the guidelines morass

UI design guidelines come in many flavors and from a variety of sources. The first task in developing corporate guidelines is to understand what types of guidelines already exist, and where corporate guidelines fit into the overall picture. One way to classify guidelines is a framework organized by context of use. The UI Design Guidelines Pyramid depicts the relationship between types of guidelines, where guidelines appearing above are derived, in part, from those appearing below (see figure 1).

At the base of the pyramid are general guidelines derived from basic and applied research performed in universities and corporate labs. Examples are familiar heuristic guidelines, such as "provide feedback," and "speak the users' language." Built upon this base are guidelines published by software vendors in support of their own products (2, 7, 11, 14). These companies create GUI development tool kits which include widgets and controls that locally incorporate principles of well-designed user interaction. An example is how Apple Macintosh incorporated direct manipulation instead of a typing dialog style in order to help "prevent error" in expressing file command syntax. Sets of these guidelines are usually referred to as "style guides." Also at this level are published guidelines (4, 5, 6, 8, 9, 10, 12, 17, 18) written by authors who are expert in HCI, drawing their recommendations from their own and others' research and development experiences.

The guidelines discussed thus far are relatively free of task context. They have to be. Vendors and researchers know only so much about the specific applications that companies want to build or the individual development tools they choose for building them. Together, these guidelines form a base kit of construction design rules—building blocks from which to choose for supporting specific design environments. However, not all these guidelines are practical in the corporate environment. With the addition of detailed information about corporate applications and development environments, comes the real-world constraints that make the implementation of certain guidelines impossible, impractical, or irrelevant.

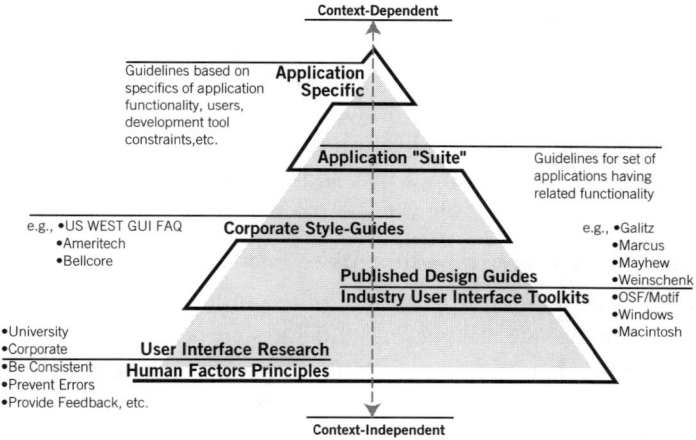

Figure 1. UI Design Guidelines Pyramid.

Corporate guidelines: incorporating context

Moving up the pyramid one more level are "corporate" guidelines. These are guidelines chosen by a company that provide design guidance applicable to their own types of applications and particular computing environments. Corporate guidelines sample from the bountiful menu of published guidelines. In addition, they resolve ambiguities when they exist, provide further clarity as needed, and provide extensions across multiple platforms where appropriate. They also include brand new guidelines needed for the corporate environment that do not appear anywhere in published form. Some companies, for instance Bellcore (3) and Ameritech (1), have created and made available their corporate design guidelines.

Continuing still higher on the pyramid are "application suite" guidelines. These are guidelines that companies may create specifically for groups (or "suites") of applications that are designed to work together within the context of related tasks performed by a targeted user population. For many application suites, it may be important to provide a consistent look and feel to give users the impression they are working with a coherent system (or set of systems). Often, responsibility for creating the applications within a suite falls to the same corporate development group. That group may decide to augment corporate guidelines with additional guidelines that meet the needs of their specific UI development tools and common application functionality.

At the top of the pyramid are "application-specific" guidelines. These address the fact that individual applications often have unique design needs driven by unique functionality and the use of particular UI development tools. An application project team may decide it is important to define guidelines that address these needs, and help drive consistent interaction for common elements within the application.

Published guidelines: impediments to use

Corporate style guides, application suite guidelines, and application-specific guidelines are all types of guidelines corporations may develop. As discussed, there is a plethora of guideline references already published. It would be a relatively easy task to implement corporate guidelines if it only involved handing out a copy of the Motif Style Guide—or at worst, a copy-and-paste exercise from a few guidelines references.

Reality is different. Useful corporate guideline material exists in a number of published sources. However, this material must be synthesized and edited. The sheer size and number of available guidelines references would quickly extinguish the motivation of the typical software developer. Similarly, sifting through the text of many of these guidelines will leave that same developer convinced that his/her own preconceived design ideas are preferable to whatever the guidelines are trying to say (even if they could understand them). Even when a developer grasps a partial understanding of a guideline, it may serve only to reinforce the notion that his/her own design ideas are somehow consistent with the guideline (even when an "expert" would say they are not).

The impediments to creating usable corporate guidelines stem from the same difficulties in communicating complex ideas that are evident in published guidelines in general. Many communication impediments can be traced back to five root causes: lack of measurability, reliance on assumed design knowledge, lack of context, confusing language, and accessibility. These are illustrated in the following sections using examples from available design guideline references.

Not measurable

An important warning sign that a guideline may miss the mark is when its application cannot be independently and objectively verified via inspection. For example, while you may agree with the following guideline, it might be very difficult for any third-party evaluator to agree that the guideline has been satisfied in a particular design.

The principle of alignment states that nothing should be placed on the page arbitrarily. Every item should have a visual connection with something else on the page.

The more subjective a guideline, the greater the variance in its interpretation and potential for misapplication. A successful guideline is measurable and produces a very high percentage of agreement by independent evaluators as to whether it is applied correctly in a particular design. However, guideline verification must also be practical.

Surfaces in the middle of the visual field should not have a brightness contrast of more than 3:1. (4, p. 322).

This guideline, while measurable, is nevertheless not a very useful one because it cannot be measured by practical means.

Reliance on assumed design knowledge

Guidelines often fail when their successful application relies on the assumption that the reader has a larger body of related design knowledge that is independent of the specific guideline itself.

Use Meaningful and Recognizable Representations... (Design tips): To provide a meaningful and highly visual representation, the application designer should consider both functional and aesthetic aspects of the user interface elements or objects. A representation of an object that is meaningful helps the user both to transfer knowledge about the real-world object to the computing environment and to remember the relationship of the object to other objects." (9, p. 9)

While this guideline may make sense to an experienced visual designer who understands concepts such as mental models, it may not carry the same explanatory power to the novice designer. This type of guideline may not give enough practical assistance to a designer/developer who has an interface design task at hand.

Lack of context

Many, if not most, guidelines should provide some additional context to the reader in order to improve communication. Context is often best provided in the form of visual examples. For example, without context, the following guideline leaves the door open to misapplication.

> "Use color sparingly. Color is a powerful attention-getting technique. Use it sparingly or it loses its effectiveness. Do not use it only for aesthetic purposes. Every time you use color, it should be for a specific, attention-getting reason." (17, p. 72)

Without providing context along with this guideline, it may be difficult to reason with a developer that their use of color for a "specific, attention-getting reason" was misguided. Another example of why contextual examples are often necessary is provided by Apple:

> "Modelessness. For the most part, try to create modeless features that allow people to do whatever they want when they want to in your application. Avoid using modes in your application because a mode typically restricts the operations that the user can perform . . . In contrast, modelessness . . . gives the user more control. . . . This is not to say you should never use modes in applications. . . ."
> (2, p. 12)

Confusing language

There is no substitute for a clearly worded guideline. A good guideline is concise and gets to the point quickly. The following guideline illustrates this point.

> "There are a number of alternatives to properly binding **BTransfer** and **BMenu**:
>
> The mouse can be treated as a 3-button mouse if chording the two buttons is treated as the third button. . . . The chorded buttons can be treated as the second or third buttons on a 3-button mouse, in which case the unchorded buttons are treated as the first and third, or first and second buttons, respectively." (14, pp. 2–5)

Just as it is important to "speak the user's language" when designing user interfaces, it is also important to use guideline language consistent with the intended user audience. Guidelines intended for software developers should not contain jargon used by HCI experts.

Accessibility

Finally, if guidelines are not easily accessible, they will not be used. Developers do not have the time or patience to wade through textbook style prose and infer their guideline discoveries. Many developers are of the feeling, "Just tell us, and we'll do it."

Likewise, developers do not want to play the part of researcher and resort to reviewing multiple sources of guidelines. In fact, multiple sources sometimes lead to confusion where guidelines seem inconsistent with each other. Consider the following guidelines on menu organization:

> Place the most frequent or critical items at the top of the pull-down menu. (17, p. 55)

> In general, place the most frequently used menu items at the top of the menu. . . . However, create groups that make sense to the user rather than simply grouping all the most-used items at the top of the menu. (2, p. 60)

> Order menu choice labels according to convention, frequency of use, order of use, categorical or functional groups, and/or alphabetical order, depending on the user and task variables. (10, p. 154)

> Therefore, it is always a good idea to provide a few concise and non-conflicting sources of guidelines. Use a standard reference style format which highlights the guideline and uses structured progressive disclosure to augment the guideline with other information deemed relevant, such as examples, rationale, implementation tips, etc. This allows users to quickly find the information they want. The use of hypertext, the Web, and corporate intranets make a terrific delivery platform for this kind of structure and universal accessibility.

Stake your CLAIM

Avoiding the pitfalls described above is key to creating usable corporate

UI design guidelines. The advice is summarized by five principles called "CLAIM" (see figure 2). Individual candidate guidelines can be reviewed against the CLAIM principles of context, language, independence, and measurability. An entire set of guidelines and their implementation can be reviewed against the principle of accessibility.

Context   Provide visual examples,
                       do's/don'ts, etc.

Language   Speak the audience's language; get
                       the point

Accessibility   Use reference-style format; progressive disclosure for
                       supplementary or related guidance

Independence Implementation doesn't rely on other assumed design
                       knowledge

Measurability Compliance can be objectively evaluated via
                       inspection or practical measurement

Figure 2. Five Principles of Usable UI Design Guidelines—CLAIM.

Summary

Creating corporate design guidelines need not be a useless exercise. These guidelines do have a place in the overall UI design picture. However, it is necessary to match your expectations to what is achievable. The mere production of corporate guidelines will not ensure understanding or usage.

It is important not to derive corporate guidelines for everything you can think of. The 80/20 rule applies appropriately to this task. Approximately 80 percent of the design questions or misapplications you see relate to about 20 percent of the design space (e.g., What colors and fonts should I use?). Therefore, begin by concentrating on the vital few guidelines that you think will keep the phone from ringing off the hook.

Keep redundancy with industry style guides to a minimum. Sometimes it makes sense to repeat this information because it increases coherency or reinforces important guidelines. As a rule, however, your corporate guidelines will demonstrate more value by addressing issues not already addressed adequately by industry style guides (and there are plenty!). This approach also makes sense from a guidelines maintenance perspective.

Also, accept the fact that written guidelines are not a good solution to many complex design issues. When this is the case, suppress the urge to create a poor guideline that is likely to be misapplied, or at best, confusing. Instead, omit the topic all together. After all, that's why we need design experts!

Finally, follow the principles of CLAIM when creating your targeted guidelines. They will help you in deciding if your prospective guidelines are good candidates for being useful and usable in the real world.

Selected guidelines readings

Ameritech.http://www.ameritech. com/news/ testtown/library/ standard/guix3.html#3.1.3.

Apple Computer, Inc. *Macintosh User Interface Guidelines.* Reading, MA: Addison-Wesley, 1992.

Bellcore (McFarland, A., Dayton, T.). *Design Guide for Multiplatform Graphical User Interfaces.* Bellcore Document No. LP-R13, December 1995.

Fowler, S., Stanwick, V. *The GUI Style Guide.* Boston: AP Professional, 1995.

Galitz, Wilbert O. *User Interface Screen Design.* New York: Wiley-QED, 1993.

Galitz, Wilbert O. *It's Time To Clean Your Windows: Designing GUIs That Work.* New York: Wiley-QED, 1994.

IBM Corporation. *Object-Oriented Interface Design*: IBM Common User Access Guidelines. Carmel, IN: QUE, 1992.

Kobara, Shiz. Visual *Design with OSF/Motif.* Reading, MA: Addison-Wesley, 1991.

Marcus, A., Smilonich, N., Thompson, L. *The Cross-GUI Handbook For Multiplatform User Interface Design.* Reading, MA: Addison-Wesley, 1995.

Mayhew, Deborah J. *Principles and Guidelines in Software User Interface Design*. Englewood Cliffs, NJ: Prentice Hall, 1992.

Microsoft Corporation. *The Windows Interface Guidelines for Software Design*. Redmond, WA: Microsoft Press, 1995.

Mullett, K., Sano, D. *Designing Visual Interfaces: Communication Oriented Techniques*. Mountain View, CA: SunSoft Press, 1995.

Nielsen, J. *Usability Engineering*. Boston: AP Professional, 1993. (*See chapter 8.)

Open Software Foundation. OSF/Motif Style Guide Revision 1.2, Englewood Cliffs, NJ: Prentice-Hall, 1993.

Tetzlaff, L., Schwartz, D. R. *"The Use Of Guidelines In Interface Design."* In CHI '91 Conference Proceedings. ACM Press, 1991, pp. 329–333.

Thovtrup, H. Nielsen, J. *"Assessing The Usability of a User Interface Standard."* In CHI '91 Conference Proceedings. ACM Press, 1991, pp. 335–341.

Weinschenk, S., Yeo, S. *Guidelines for Enterprise-Wide GUI Design*. New York: John Wiley, 1995.

Williams, R. *The Non-Designer's Design Book: Design and Typographic Principles for the Visual Novice*. Berkeley, CA: Peachpit Press, 1994.

UPA '99 Workshop Summary: Integrating Human Factors Analysis Methods with Use Cases

Daniel Engelberg,
Computer Research
Institute of Montreal

Abstract

In this article, we summarize a UPA workshop in which we looked at the analysis tools of Hierarchical Task Analysis (HTA) and use cases, and asked how they could be integrated. The reasons for considering an integration are twofold: first, the information provided by HTA is not complete for interaction designers, and second, this valuable technique tends to be ignored by developers who are doing user-interface (UI) design. We concluded that what we need is a new conceptual framework, supported by a computerized tool that provides different views of the same underlying task data.

Introduction

The issue we started out with seemed like a relatively straightforward question: How can Hierarchical Task Analysis (HTA) and use case methods complement each other? They are both methods for describing a task, but they seem to convey different information. HTA is a classical technique in HCI but rarely used by developers (non HCI specialists) who are doing UI design; use cases are a valuable tool for developers but are often ignored by HCI specialists.

The purposes of hierarchical task analysis (HTA) and use cases

The purpose of HTA is to understand the *why* of the task at the same time as the how.

The purpose of use cases is to provide immediate and concrete design and development information to the developer, i.e., information for designing the underlying system. Although the technique was developed primarily for system design, it was transferred to GUI design when GUIs started becoming popular.

Example 1: Definition and example of Hierarchical Task Analysis

Hierarchical Task Analysis (HTA) is a task representation in which the task is decomposed progressively from a top-level goal to intermediate and then lower-level goals and methods. The representation forms a tree or document outline format. If we consider all of the immediate sub-goals of a given goal, these sub-goals are listed in the order in which they are reached. Various operators can identify conditional and temporal relationships among goals.

Sample Hierarchical Task Analysis:

1. Handle social intervention requests
   1.1. Receive request
   1.1.1. Get details of the request
   1.1.2. Check if there is an existing file for the client
   1.1.3. Consult the client's file (if one exists)
   1.1.4. Respond to an immediate request
   1.2. Evaluate the request
   1.2.1. Clarify the request
   1.2.2. Globally evaluate the request (optional)
   1.2.2.1. Understand the client's situation and current problem
   1.2.2.2. Determine the needs of the client regarding this problem
   1.2.2.3. Give an opinion on the problem; judge the client's motivation
   1.2.3. Evaluate the urgency of the situation
   1.2.4. Identify solutions
   1.2.5. Consult manuals of practice (optional)
   1.3. Arrange for treatment or referral

Differences between HTA and use cases

The primary driving difference between the two approaches is that HTA is an attempt to understand the user's task largely independent of the system, whereas use cases document an interaction that a user has with the system. See the HTA illustration in Example 1.

HTA is a powerful tool for representing a task at a conceptual and teleological level and revealing its sequence of operations. However, HTA doesn't encourage documentation of idiosyncratic strategies of use: It tends toward generalization and abstraction. Indeed, it is precisely this abstraction that gives HTA its strength. In addition, HTA makes it easier to follow the flow of events because they are presented in a more connected way than with use cases.

The strength of use cases in system design is clearly their ability to represent specific scenarios in detail. However, they are not optimal for UI design, because, at least in practice, they fail to provide an overview of the task, and the task-related information is often obscured by system-level detail. See the use case illustration in Example 2.

Use cases are like snapshots of parts of a task. It is possible to put them next to each other but it's still difficult to get the whole picture and understand how the parts are connected. Developers sometimes "boil down" their use cases to a more conceptual level, but these use cases still lack a description of the overall sequence and structure of events.

Despite these differences, use cases are not irreconcilable with HTA. Primarily, they provide more detailed information, and they provide information on dimensions of the task that are not covered by typical HTA approaches. They usually involve less abstraction and generalization; they are a more concrete presentation of the facts.

The users of HTA and use cases

Several different participants in the UI design process need task information: human factors analysts and designers, expert users working as analysts and designers, developers doing UI design, developers doing system design, designers, and managers. These groups have different needs and objectives.

HTA is especially responsive to the needs of human factors specialists. HTA can also be useful to expert users who are performing analyses on their own tasks, helping them think more deeply and consciously about their work. In general, however, expert users have less need for abstract representations of the task, and certain groups of expert users may have difficulty relating to the technique.

In principle, HTA can and should help developers at the UI design phase, but in practice, the approach doesn't seem to appeal to them. We'll discuss this issue in more detail later.

Conversely, use cases are useful for developers doing both UI design and system design, because they present concrete information about the task in an immediately accessible manner that does not require any interpretation. They also help developers, because they give extra context about the task, which helps the developer reconstruct the sequence of activities.

Different users of task information need different formats

The discussion of the needs of different users of task information led us to understand that HTA and use cases have different objectives and uses adapted to their respective users. The problem is that although they are adapted to their respective groups of users, neither provides complete information in itself for the design and programming of the interface. The respective groups treat the tools as if they provide sufficient information, but each representation has its own gaps and weaknesses.

The ideal approach would be a single tool that integrates the different models

To meet the information needs of different groups of users and the tasks of programming and design, and to allow for coherent communication between the two groups, the ideal approach would be a single framework that integrates the different representational models of the task.

This integrated framework should have the following properties:

- It should provide more complete information about the task for all users of the information.

- It should encourage developers to use HTA information and encourage human factors people to use more complete and concrete use-case information.

- It should provide a flexible and multiple representation, i.e., different views of a central database of information.

- It should establish and maintain coherence between the different formats (and contents) of representation.

- If possible, it should help users understand the relationships between the different formats (and contents) of information.

The solution that presented itself was a computerized tool that knits together the different types of information and allows users to take different views of the same underlying model: an HTA view, a use case view, and a story view that allows users to see the task as a linear sequential activity from start to end. The framework could be extended

Example 2: Definition and example of a use case

A use case is a collection of possible scenarios between the system under discussion and external actors, characterized by the goal the primary actor has toward the system's declared responsibilities and showing how the primary actor's goal might be achieved or might fail.

Sample use case

Use Case: Evaluate Urgency of the Request

Characteristic Information

Goal in Context: Social Worker evaluates the urgency of the request
Scope: Health clinic psycho/social
Level: Sub-function
Preconditions: Client has been identified
Success End Condition: Request has been classified as urgent or not urgent
Failed End Condition: Urgency not identified
Primary Actor: Social Worker
Secondary Actors: Requester
Trigger: Request comes in

Main Success Scenario

(The following are done in parallel, without any predefined order)

• Social Worker checks if the Client is suicidal
• Social Worker identifies the Client's support
• Social Worker evaluates the Client's vulnerability
• Social Worker evaluates the Requester and/or the Client's perception of urgency

Extensions

Sub-variations

Related Information

Priority: Top
Performance Target: 2 minutes for most requests
Frequency: 20/day
Superordinate Use Case: Get Details of Request
Subordinate Use Cases: None

to include other types of models such as the data model and the business process model.

Tool properties for getting developers to use HTA

In general, developers may prefer use cases because they are more restricted chunks of information and easier to deal with for local design issues. Developers are generally not designing for the global design at a given moment. When they are programming they need concrete information. In general their task as programmers is not to stop and think, but rather to act.

Thus, the tool needs to have the following properties:

- The translation (relationship) between use cases and HTA must be clear.

- The framework must recommend the appropriate use of use cases.

Representations (views) to be included in the tool

Although the workshop covered only use cases and task analysis, in the long term, the tool and framework that we are proposing would ideally support views of all of the following representations of the task and the relationships among them:

- Use case

- Task analysis

- User's model of the task

- User's model of the interface

- Designer's or developer's model of the interface

- Data model

- Business process model

Ultimately, this set of representations could also be linked with the design model or functional specifications.

Conclusion

The number of different models and objectives and technical cultures in the field of computers makes communication difficult among the groups involved in design and development, especially the designers and developers themselves. These communication difficulties can lead to design problems.

What we propose is a new conceptual framework for handling the different models used in a design and development project, combined with an application tool to support the framework. The framework and tool that we are proposing would provide different views of an integrated central database of information and assistance on when and how to use the different models. The different views would not necessarily be equivalent translations of exactly the same set of information, but they would share a common core of information.

Such a tool would help correct the present overemphasis on the differences among the models and provide the much more complete and dynamic information set required for design. The tool would allow a more integrated design, one that is based on more complete information.

Ultimately, this tool could become the front end to a GUI design application.

References

Ivar Jacobson, 1994. Object-Oriented Software Engineering: *A Use Case Driven Approach.* New York: Addison-Wesley Object Technology Series.

Ivar Jacobson, Grady Booch and James Rumbauch, 1999. *The Unified Software Development Process.* New York: Addison-Wesley Longman.

The Third Dimension in Paper Prototypes

Simo Sade and
Katja Battarbee
University of Art and Design

Low-fidelity (lo-fi) prototyping must be easy and cheap to be practical to use. One method for lo-fi prototyping of user interfaces is paper prototyping, but paper prototyping cannot fully represent products that involve physical manipulation. The level of representation can be improved with three-dimensional paper prototyping ("3DPP"). Designing smart products such as mobile phones, VCRs, or ATMs can benefit from 3DPP because their interfaces consist of physical shapes, buttons, and small displays, and using them may also involve handling physical objects such as money, receipts, or cassettes.

Smart products are often not very usable. The many reasons cited in the literature are related to three issues: the technology-driven design, the problem of operating everything with only a few buttons and a small display, and the heterogeneous user group. According to Shackel (1991), the overall acceptance of products depends on their utility, usability, and their likeability, balanced against their financial, social, and organizational costs. Smart products, which are often consumer products and used casually or for fun, need to be made easy to use in their actual context of use.

Smart products should be understood as "whole-product user interfaces" (Vertelney & Booker, 1990). In evaluating the usability and likeability of smart products, the wholeness of a product in context should be considered. The wholeness consists of the physical qualities, shape and size, the interaction, and the context of use.

How can the wholeness of the smart product be represented early in the design process? Shape and size are traditionally represented with sketches, drawings, and CAD models. The tangible qualities can be represented by 3-D models. User interfaces can be represented using user interface maps, paper prototypes, and interactive on-screen prototypes. A 3-D paper prototype is an attempt to represent the whole product—shape, size, and user interface. We applied 3DPP to two product development cases: an aluminum-can refund machine and a mobile computer for in-vehicle use. These cases are described below.

An early example of the 3DPP technique

First, however, to illustrate and validate the 3DPP technique, we created a three-dimensional paper prototype of an imaginary product. As an example, we came up with a smart card terminal to be used in public places. The terminal would be able to receive and send faxes and print and send e-mails. 3DPP is well suited to designing products of this type. Using the device involves both operating the software user interface and physical manipulation: the user inserts a card, opens the device to use it, and presses buttons, reads screens, opens scanner lids, and adds paper.

In using the 3DPP to evaluate the design idea with users, the facilitator changes the printed screen cards on the prototype to match to the user's actions.

The smart product example demonstrated that we could use 3DPP to test the early designs of smart products.

Figure 1. Changing the screen card.

Figure 2. Adding more paper.

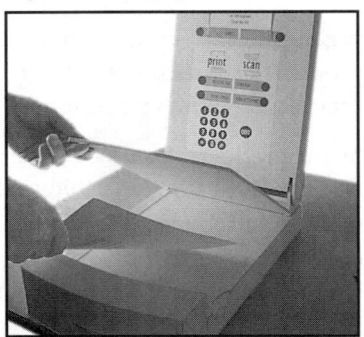

Figure 3. Scanning a document.

Case study with 3DPP: the aluminum can refund machine

Halton Systems Oy is a Finnish manufacturer and supplier of bottle and can refund machines and systems to shops and supermarkets worldwide. In 1996, they started a product development project to design a new aluminum-can refund machine. The following case study is described in Sade, et al. (1998) in more detail.

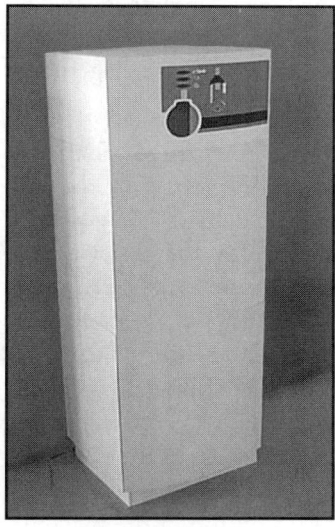

Figure 4. Mock-up of an aluminum can refund machine.

The aluminum can refund machine takes the can, checks for 100 percent aluminum content, and reads the bar code to see if it is refundable. The cans are then crushed, collected, and counted, and the user gets the refund either as a receipt to be cashed or as coins. Refund machines are typical walk-up-and-use products. We evaluated two concept alternatives in which the only difference from the user's point of view was how the user fed the cans into the machine. In creating the prototypes, we did not use any text on either interface, only three lights: red for "remove the can," yellow for "non-refundable," and green for "can refunded."

With the "automatic" machine, the users put the cans through a rectangular opening onto a tray, which then rotates the cans into the machine. The "manual" alternative has a round opening; the users have to position the can bottom first so that the bar code is on top where the machine can read it. The manually operated alternative was expected to be cheaper to produce. However, we suspected that the manual concept might be too difficult to use.

To decide between the automatic and manual models, we tested the two prototypes with real users very early in the process, before any hardware or software had been built. This early testing was a new experience for the company.

The test

Two full-size mock-ups were built of sheets of Fome-cor®. They were white, and the front panel graphics were color printed and glued on.

The indicator lights were simulated by pinning on pieces of colored paper. The designer, who parodied the "Wizard of Oz," stood behind the mock-up with her hands inside the machine, receiving or refusing the inserted cans. Another designer operated the lights on the front panel of the machine, standing next to the user.

First, an informal pilot test was conducted with two users. Then, the actual test took place in a large supermarket. Altogether, twenty customers volunteered, having been told about the test and its goals. They had to pretend they were realistically returning the bag full of aluminum cans and feeding them into the machine.

The 3DPP technique proved useful for the test. Users forgot about the cardboard box and the designers, and became immersed in the task activity. They talked to themselves, saying things such as, "Why won't it take this?" Valuable information was gathered for the development project: Users found the automatic feeder concept clearly easier. As expected, the manual positioning of the can and its bar code were difficult. The tests also revealed that only about one-third of the users paid attention to the indicator lights. Instead, they looked down at the bag to pick up the next can. A real light would get more attention, but that problem was counterbalanced by the intrusiveness of the designer's hand and actions of changing the colored papers. Another unexpected outcome was that the users put the refused cans on top of the machine.

The results from the 3DPP evaluation led to a design decision to use the more expensive but easier-to-use automatic version. The indicator lights were not implemented because users had ignored them, and they seemed unnecessary. The early test procedure was also used in marketing the new product to large institutional clients before launching the product. The test served as a proof of quality.

Later, a functioning, full prototype was built and placed in the same supermarket. After observing fifty users, we found that our usability conclusions were very similar to those we obtained with 3DPP.

Case study: a mobile computer for in-vehicle use

Aplicom Oy is a manufacturer of computers to be used in vehicles. Together with the design consultancy E & D Design Oy, we conducted a user test with an in-vehicle computer, which was intended to be used by courier or taxi companies to communicate with the drivers. The computer receives messages from the central dispatcher. The driver selects jobs, writes off finished tasks, and sends other kinds of messages using the computer. The test users were two car couriers, and the test environment was a van in a carpark. Couriers are supposed to use in-vehicle computers only when parked, although they sometimes used them while driving. Both users had experience with an earlier, text-based device. The new device had a graphic interface, soft keys, and a remote control, which allowed the device to be placed more easily inside the passenger compartment.

The model was high-fidelity, made by a designer. The screens were printouts placed on the model, which was attached to the fixture for the old device in the van. The basic tasks—get a delivery job, accept it, sign off collection, and sign off delivery—were evaluated in the van using 3dpp, and additional tasks—messages, settings, etc.—were evaluated indoors with an on-screen simulation on a computer, operated with a mouse. The sessions were videotaped.

Figure 5. The test setting in the real-use environment of the in-vehicle computer.

The aim was to get a rough idea of the usability in the real setting and to obtain feedback on the design, ergonomics, and input methods. The test and the interview were informal, and the tasks were presented using task scenarios. Background information, such as computer experience, was collected during the interview. The test revealed that the physical setup was good but there were some difficulties with the graphical interface of the device, and some icons were not self-evident. The driver with more previous Windows experience did better with the graphic interface. The test also revealed that in the real user environment, one driver used the remote, but the other one did not. We believe that showing users physical mock-ups or evaluating 2-D paper prototypes does not have the power to reveal this kind of information about physical manipulation of the device.

The difference between the 3DPP and the on-screen simulation was that in 3DPP, physical buttons, and display items are naturally different and cause no confusion, whereas in the screen simulation, they tend to get confused because they all are pictures on the screen. The couriers attempted to click the display on the screen simulation although only the buttons were intended to be clickable.

Conclusion

3DPP prototypes combine methods of user-interface design and industrial design. They represent the basic functionality and the basic design ideas in 3-D and allow the evaluation of the cognitive and physical parts of the interaction as well as the design ideas. With smart products, where the interface is as much in the overall physical shape of the product as in the screen, involving the whole interaction is essential to evaluating the real usability of the product. The real-use context can also be understood best when the 3DPP prototypes are evaluated in the field. Since many smart products are also consumer products, evaluating the acceptability of the design idea is important. The results are information for the further development of the design.

Paper prototypes in general have certain advantages over higher-fidelity prototypes (see, for example, Rudd, et al, [1996]). They are cheap and fast to build. They can be used for gathering requirements, evaluating multiple concepts, and testing before any real investments have been made in software or hardware. Screen simulations may look too final and guide attention to details instead of general issues. Screen simulations of products also confuse the user when both physical buttons and on-screens objects are portrayed as pictures on the screen and the naturalness of the physical interaction is lost.

The results with paper prototypes are not quite as reliable as with high-fidelity prototypes. Some types of interaction are difficult to simulate on paper, such as text feed, fast changes, and large interface structures.

Three-dimensional prototypes share the shortcomings of paper prototypes in general. Building them also requires a little more skill and work. But, unlike 2-D paper prototypes, they let us study the product design solutions, the physical ergonomics, and the influence of the user's physical activities on the usability of the product. Two-dimensional paper prototypes can be used to represent 3-D devices when all the controls are on one face, not much physical interaction is required, and a separate model can be used to evaluate the design idea. Three-dimensional paper prototypes have benefits when the product is evaluated for its design, when the product is a redesign of a physical product, when the operation of the device requires physical actions, or when the evaluation should be conducted in the real environment. In the early stages of the design process, 3DPP gets one step closer to representing the "whole-product user interface."

Acknowledgements

We want to thank Sirpa Riihiaho and Marko Nieminen at the Helsinki University of Technology and Heikki Salo and Kirsi Svard at E & D Design Oy for cooperation in the case studies. This study has been partly funded by the Academy of Finland and the Technology Development Centre Finland.

References

Rudd, J., Stern, K., and Isensee, S. Low vs. high-fidelity prototyping debate (1996).

Interactions, 3, 1 (Jan. 1996), pp. 76–85.

Säde, S., Nieminen, M. and Riihiaho, S. "Testing Usability with 3-D Paper Prototypes—Case Halton System." *Applied Ergonomics.* Vol. 29, No 1, February 1998. Special issue on consumer products.

Shackel, B. "Usability—context, framework, design and evaluation." Shackel, B. and Richardson, S. (Eds.). *Human Factors for Informatics Usability.* Cambridge University Press, Cambridge (1991).

Vertelney, L. and Booker, S. "Designing the whole-product user interface." Laurel, B. (Ed.), *The Art of Human-Computer Interface Design,* Addison-Wesley, Reading, MA (1990).

Conceptual Modeling

Building the Conceptual Model and Metaphor: the "3x3"

Carol Righi, IBM
Usability Engineering

Introduction

In this paper, I will discuss a methodology for helping to build a conceptual model and metaphor, a key element of interface design. The "3x3" gives interface designers a way of seeing the conceptual model through the eyes of the user. I first define a conceptual model, explain its importance, and provide several examples. Then I discuss the 3x3—what it is, and how to use it to help build a conceptual model and select a metaphor that will meet users' task needs while being easy to use and pleasant to experience.

What is a conceptual model?

The term "model" is used in many different ways by cognitive psychologists, software designers, and human factors engineers. Let's first distinguish a "mental model" from a "conceptual model." A mental model is a conception of how the world works, the way it's structured. Everyone has mental models; they help us understand and predict the behavior of new events by processing them in terms of existing concepts.

While a mental model is individual, a conceptual model is shared. The conceptual model of an application or solution is created by designers and surfaced to users via the interface. The conceptual model consists of the objects it contains (and by implication, the tasks it supports); their behaviors; and the relationships among these objects. Integral to the model is its metaphor, which is intended to communicate the nature of the model to users by comparing it to something familiar to them.

Why be concerned with mental models and conceptual models?

Conceptual models, by tapping into users' mental models, can make a system easier to use. If a user encounters a new interface that looks like something familiar, the user will already know what to do, and will have expectations about how the system will respond. Therefore, the goal for an interface designer is to build a conceptual model that taps into existing mental models.

A classic example of a conceptual model and its metaphor is the desktop. The desktop represents the conceptual model of the operating system. It takes advantage of users' experiences with using a desktop in an office environment (their existing mental model). The desktop model helps make the system easier to use because a user already knows how to interact with a desktop.

The objects in this conceptual model are represented as metaphors related to the overriding desktop metaphor: To write a letter, you access writing tools. To discard something, you throw it in the trash. To organize documents, you put them in folders. Building a conceptual model for the operating system that employs a common overriding metaphor and associated metaphors makes this system easier to use.

The "3x3" is a tool that has been traditionally used by advertising agencies and multimedia developers, among others. In a traditional 3x3, a designer or design team will generate three high-fidelity alternatives of a solution. They'll produce each to a depth of three levels. For example, in a multimedia application, a designer will develop screens that represent the first three screens of a user's path. In an advertising brochure, a designer might develop the cover image and two internal images of a brochure. These 3x3 "studies" are then shown to the client for approval.

However, the 3x3 that is used by interface designers has a different nature and goal. Rather than to obtain client approval, the 3x3 is used to explore conceptual models and metaphors with users. In this type of 3x3, three alternative models are developed as low-fidelity prototypes (hand-drawn sketches). Each is developed to a level of three-screens-deep. Then these proposed models are shown to users. Users work through a couple of common tasks while thinking aloud. The models are tested to determine if the users can understand the metaphor, and whether they support users in the completion of their tasks. Once a model is selected, a second round of the 3x3 is undertaken. The model is rendered as higher-fidelity screens to determine which visual treatment is most appealing, most appropriate, and most supportive of users' tasks.

The main advantage to implementing the 3x3 early in the design of a product is to allow alternative solutions to be explored with users prior to expending precious development resources. Unfortunately, under the pressures of time, a single solution is frequently chosen and driven forward without having validated that it is the best possible solution. But it is equally important to use the 3x3 only when necessary prerequisite design information becomes available. Prior to the 3x3, you must have gathered user tasks and requirements from representatives of your intended audience. These will dictate the content, structure, and organization of the proposed models, and will suggest appropriate metaphors. Also, if appropriate, you should have done a competitor evaluation. Competitor information allows you to determine which models will be competitive once they are developed. The design team then, using these data, engages in a brainstorming session. Model choices are considered in light of any constraints, and are pared down to three. At this point, a 3x3 is done.

Case study: an automotive kiosk

The following describes the steps followed in the design of an automotive kiosk. The kiosk was to be housed in a mall setting, and its content was intended to provide information about vehicles manufactured by the client. The steps preceding the 3x3 will also be discussed to provide the context for understanding how and when the 3x3 was used.

1. Gather user tasks and requirements

The first step in our design process was to gather tasks and requirements from the target audience for the product. In this case, we held two sessions using our Decision Support Center, an electronic meeting room. We recruited about 30 participants from two market segments. We gathered information about the tasks these participants typically perform when shopping for a new vehicle.

We gathered the details associated with these tasks, what makes them easy, what makes them difficult, plus triggers, outcomes, etc. We also had the participants group these tasks into logical categories using an affinity grouping exercise.

2. Gather competitor information

While we were collecting task data, we also collected competitor data. We targeted two sets of kiosks as competitors: automotive kiosks and other best-of-breed kiosks in general. For the automotive kiosk evaluation, we attended the annual Detroit Auto Show, at which about ten kiosks were examined. For the best-of-breed kiosk evaluation, we visited the EPCOT center in Orlando which houses dozens of kiosks. At each venue, the kiosks were examined heuristically, with regard to strengths and weaknesses, use of models and metaphors, and overall appeal with regard to providing a *total user experience.*

3. Brainstorm alternative conceptual models, metaphors

With marketing-, task-, and competitor-data in hand, the design team engaged in a brainstorming session. The goal of this session was to propose a set of models and metaphors for our kiosk design. The team came up with about 12 models, including both concrete models (e.g., the showroom, the road), and some rather abstract models (e.g., the elevator, the board game). As we brainstormed, we talked through what the design might look like, for example, "Each square on the game board could represent a step in the vehicle-shopping process. A user can select a game piece, which can be one of several types of vehicles. . . ." and so forth. As we brainstormed, we tapped the existing data to be sure the alternatives we were proposing could be designed to support the tasks, wants, and needs of the target market; and would prove to be competitive with existing solutions. This step also helped us determine where a metaphor worked, and where it didn't. In our board game model, for example, we found that using dice or a spinner to move a game piece broke the conceptual model:

Dice and spinners imply randomness. Our task analysis indicated that vehicle-shopping is performed rather methodically by most of our participants. Through this process, we refined and eliminated several models.

4. Weed out/Pare down models

After talking through several possible models, we revisited each, this time keeping in mind several possible constraints to the design. We considered the technical feasibility of these models: Could they be implemented on schedule? At this early stage of design, it is difficult to gauge implementation needs; however, a rough "feel" for implementation resource needs is possible. Then, we considered whether each model could be designed to be consistent with the existing marketing materials of the corporation. Such consistency is desirable, as it reinforces a corporate image, contributes to brand recognition, and gives users a sense that a single team designed all aspects of the vehicle-shopping experience. Finally, we talked through each remaining model a bit further. More of our original models were thereby eliminated.

5. 3x3 Phase I

As a result of the brainstorming and weeding/paring, we selected our three best candidates for the 3x3. We tried to include a mix of abstract and concrete metaphors. A visual designer sketched the first three screens for each model. One model used the abstract metaphor of an elevator (figure 1). The act of vehicle-shopping starts by entering one of two elevators, which brings its occupants to various floors of a showroom. Two elevators were employed to allow users to shop either by brand or by vehicle type.

We recruited three pairs of participants and engaged them in prototype evaluation sessions. We showed the designs to them and had them work through the task of getting started at the kiosk, using a specific vehicle as an example. We watched and listened as they worked through the models. We attended primarily to how well the model and its metaphors "fit" their expectations: Were they meaningful? Did they enhance or detract from the usability of the screens? Real-time "tweaks" to the designs were made in an effort to strengthen each model. The overall goal of this phase was to narrow the choices of model representations to one.

Phase I of the 3x3 resulted in our selecting a single model to further explore: the "road." Note that this model was at first rejected by the design team for competitive reasons: We found that many automotive manufacturers use the road in their marketing materials. However, the users themselves suggested this model. We therefore sketched and included it. It turned out to be the best-received model of the three tested.

6. 3x3 Phase II

We then engaged two visual designers to create three visual treatments of this model. We sought their guidance in suggesting which types of treatments might be appealing. They created one "cartoonish" treatment, one "abstract" treatment, and one "photorealistic" treatment. They created three screens for each, similar to what was done in Phase 1. We then showed these three treatments to users. The focus this time, however, was not on the model itself, but on the visual implementation of the model. We focused on whether the treatment supported the model; for example, were the screen elements and their significance in the photorealistic treatment easier to comprehend than those in the abstract treatment? We also focused heavily on overall aesthetic appeal—that is, which of the three treatments was most pleasing to participants. The photorealistic road model ultimately was most users' preferred choice.

Some methodological considerations

The 3x3 is facilitated and can be supplemented by other user-feedback methodologies. User comments are gathered through interview and think-aloud, as in a typical prototype evaluation. Single users or pairs of

Figure 1: Elevator metaphor.

These three images comprise one of three conceptual models/metaphors tested in a 3x3. Representative users were given a scenario and a task to complete. In this case, the task was to use this kiosk to find information about a certain mid-sized car. The users attempted to complete this task using this solution and two alternative solutions.

users can be employed. Designs can be presented as complete, or as simpler frameworks for participatory design sessions. Multiple designs can be shown to each user (within-subjects experimental design), or each set of users can respond to a single design (between-subjects or between-groups experimental design). The World Wide Web can also be used to gather large "n" feedback, for example, about the aesthetic appeal of the visual treatments. As with all usability engineering, the methodology should be appropriate to the situation.

The number of models and their metaphors presented in either Phase I or Phase II can also be altered. Three is a good "rule of thumb" number, but need may dictate using fewer or more models. If users suggest a metaphor other than those shown in Phase I, that model can be sketched and presented. On the other hand, if none of the models in Phase I proves desirable, it may be necessary to do a second brainstorming session. It is also advisable to present a range of concrete and abstract metaphors in Phase I to get a sense of whether users want a more accessible, straightforward representation of their tasks, or if they're willing or prefer to work through a more abstract approach. If the model proves appealing, but the treatments do not, then revisiting Phase II may be necessary. In short, you can alter the 3x3 to meet your needs.

Summary

The 3x3 gives interface designers a way of designing the conceptual model by exploring alternative models and metaphors through the eyes of their intended users. Rather than selecting a single solution and driving it toward implementation without first validating it with users, the 3x3 allows designers to explore alternatives early in the design of the interface. By focusing first on selecting from alternative models and metaphors, and second on alternatives for their visual implementation, the designer can explore whether an intended solution will meet users' task needs while being easy to use and pleasant to experience.

The author would like to hear about your experiences using the 3x3. Please write to Carol Righi at righi@us.ibm.com.

Some readings on conceptual models and metaphors:

Collins, D. (1995). *Designing Object-oriented User Interfaces.* Redwood City, CA: Benjamin/Cummings.

Mandel, T. (1997). *The Elements of User Interface Design.* New York: Wiley.

A Pattern Supported Approach to User Interface Design Process

Åsa Granlund, Ericsson
Daniel Lafreniere, GESPRO

Introduction

The use of patterns for capturing and transmitting User interface (UI) design knowledge has become a hot topic over the last couple of years. In recognition of this, a workshop on patterns was organized during the 1999 UPA conference in Scottsdale, Arizona.

As a basis for the workshop, we used our Pattern Supported Approach (PSA) to the User Interface Design Process (Granlund & Lafrenière, 1999) which served as a common ground for the participants and the basis for the discussion of a number of questions.

In this synopsis of the workshop, we first explain what we mean by patterns and then present the discussion questions and a summary of the participants' responses.

What are patterns?

A pattern is a way of describing a solution to a design problem. The pattern concept originates with architecture and was introduced by Christopher Alexander in the mid-'70s. Simply put, Alexander noticed that for a recurring problem, certain solutions would apply. He then coined the term "pattern" to describe this combination of a recurring problem/solution.

In our view, patterns are a nice way of capturing the knowledge of good designs, thus allowing designers to transmit and reuse this knowledge.

The goal of using patterns is to create an inventory of solutions to help designers (and usability engineers) to resolve UI development problems that are common, difficult, and frequently encountered (adapted from Loureiro & Plummer, 1999).

Patterns have been used in *Object-Oriented Programming and Data Modeling* (see Fowler, 1997) and (Lafrenière, 1998) over a number of years, but have quite recently entered the world of user-interface design—a first workshop on this topic was held during CHI '97 in Atlanta (Erickson, 1997). Here is an example of a design pattern taken from Granlund & Lafrenière:

Name:	Timeline
Examples:	Calendar, curriculum vitae, agenda, Gantt chart
Context:	There is a need to convey a lot of time-related composite information that may be interrelated.
Problem:	How should the information be displayed to the user?
Forces:	**Task**
	- Data analysis
	- No implicit workflow
	- Overview needed
	- Need for pattern detection
	- Need to correlate data
	- Need to compare data
	- Details needed (information drill down)
	- Information has to be easily interpreted with accuracy
	User
	- Domain professional
	- May be computer-illiterate
	Context
	- Limited time
	- Stressful environment
	- Need for accuracy
Solution:	Implement timeline into which you group related data vertically, enabling comparison between groups of data, supplying detail on demand, displaying events and episodes on the horizontal axis. Offer zooming, filtering, emphasizing, and searching capabilities.

Let us explain this pattern. It has a *name*, ideally a significant one that captures the intention of the pattern! It provides some *examples* to illustrate the concept. There is a *context* description which usually is one sentence describing the goal of the design, and there is the *problem* we are trying to resolve. Then, there are a number of *forces*. The forces describe all the factors that might influence design, and they may well be in conflict with one another. This means that the emerging solution must often provide the best trade-off between conflicting factors. The factors emerge from information gathered during task analysis: the user, the context, and the task. Note that forces can change from one kind of pattern to another.

Finally, there is a *solution* to the problem, with an illustration when it is appropriate. We want to point out that for our work with patterns for interaction design, we do *not* propose that these solutions are absolute. Rather, we see them as strong examples of solutions that have proved to be good, implicitly encouraging questioning and interpretation on behalf of the interface designer.

The terms used in this pattern description (*Name, Examples, Context, Forces, Solution*) come from Alexander's pattern notation (also called the Alexandrian notation). There is more than one notation for describing patterns, but we chose this one because we appreciated its structure both for capturing and transmitting knowledge in a simple and effective way.

Please note that the pattern in the above example is what we refer to as a *design* pattern. Our work includes different kinds of patterns, and we will talk more about these later in this report.

Why are patterns interesting?

Patterns are a hot topic today, and there are many reasons for this interest:

- Patterns provide a **lingua franca** (Erickson) that can be read and understood by all, regardless of background.

- The existing formal ways of documenting UI design knowledge are often weak—patterns offer a good way of capturing and transmitting this knowledge; they are presented consistently, are easy to read, and provide background reasoning. The format provides information about the problem at hand, the context, a solution, and the rationale behind this solution.

• They promote reuse.

Patterns are used implicitly by many skilled UI designers who have found solutions that have worked for them in the past. However, these designers usually keep little in the way of formal (or documented) descriptions of these solutions. Thus, there are in fact both *implicit* patterns and formal (or *explicit*) patterns. Explicit patterns can be used as a means of collecting and *formalizing* design knowledge.

Using patterns should be like asking your experienced colleague in the next room for advice. Patterns are a valuable source of information, supporting both the analysis and the current situation and the design of the new system.

PSA as a common ground

During the workshop, we used the pattern supported approach (PSA) to the user-interface design process as a starting point for the exploration and application of patterns. PSA addresses patterns not only at the design phase but before design. For instance, we present patterns for system definition, task analysis, and conceptual design.

The idea behind PSA is to provide a chain of patterns that support each step of the design process (patterns supporting one step point to patterns supporting the consecutive steps). These patterns offer a way to capture and communicate knowledge from previous designs (including knowledge from system definition, task/user analysis, and conceptual design) and act as tools to guide and support the user-interface designer's tasks.

Patterns **cannot** serve as a single source of design knowledge. They must still be complemented by traditional sources of information, but they will point to information that is generally valid (for a specific domain) and also to designs that have proved good for similar projects. Since the description format provides reasoning and motivation, the patterns' relevance for the current project can be tried and evaluated.

In the early phases of the design process, information patterns are used for defining the scope of the system and supplying the process with knowledge regarding task, user, and context of use. Later in the process, and based on the information patterns, other patterns suggest solutions for the system's conceptual design and screen design.

Using patterns to gain knowledge from previous designs does not require special skills, since patterns should by nature be easy to read and understand.

Revising and adding new patterns does not really require any formal training either, but necessitates practical experience of discovering and documenting patterns. So, to use patterns in the user-interface design process, some basic training and facilitation by pattern-literate designers is required.

Workshop discussions

The following provides an overview of the questions and discussions that were raised during the workshop day.

Information patterns

Does everything have to be described through patterns (specifically business domains and processes)?

Looking at traditional design patterns (which provide a solution), it is relevant to ask whether *background information* is suited to be expressed

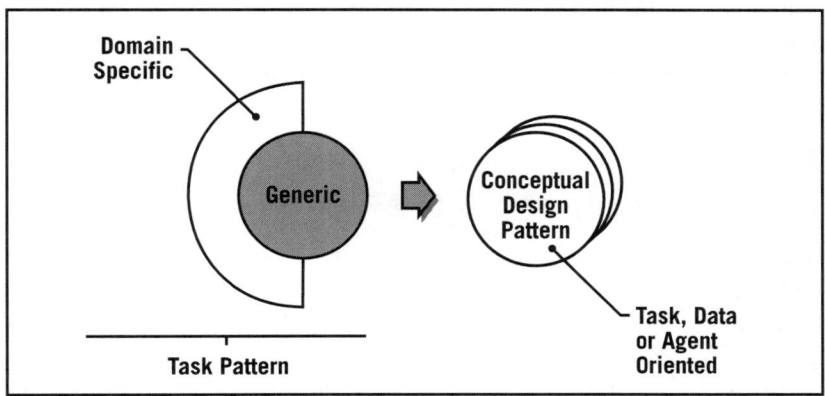

Figure 1: The PSA framework.

in a pattern format. This question was raised a couple of times during the day, but in the end we agreed that it is worthwhile to capture information using patterns. We now explicitly call them *information patterns*.

The information that we base designs on (i.e., information regarding user, context, and task information) is indeed recurring within a domain—there are patterns in information—and the Alexandrian form of capturing this information is feasible. A problem and a solution will not be included in the information pattern, but the use of context and forces is a good way of describing the information, and for pointing to the next step in the process. The distinction between information patterns and design patterns is, however, very important.

Conceptual design patterns

How can patterns contribute to the conceptual design of complex systems?

This problem is not unique to the pattern supported approach. The organization of information and tasks for a system is done at several levels of detail, and concentrating on parts of the system may cause under-optimization, since the overall structure (and the dynamics between different parts of the system) may be disregarded. Some kind of *blueprint patterns* for the overall conceptual design was suggested and will certainly be looked into further by the workshop organizers.

Another question that was raised concerned the notation to use for exemplifying a conceptual design pattern: Participants used different notations, formal and informal (mind maps, tables, modified version of entity-relationship diagram, etc.) . . . we did not have time to conclude the discussions during the workshop, but we will continue to explore this issue.

Patterns discovery or invention

Do we really discover patterns, or do we invent them?

Traditionally, it is said that patterns are not invented (designed), but rather discovered. The very thing that leads to a pattern is a recurring phenomenon in our world (a recurring design problem with a generically good solution, or as included in the PSA approach—recurring information that we keep using for our designs). This implies that all we have to do in order to discover patterns is to take a look around us.

However, the very content of the workshop implies a more organized viewpoint on patterns. Dealing with the contents, structure, and relationships among patterns makes patterns come across like any other theoretical model.

As one participant pointed out, the concept of discovering patterns is kind of a metaphor in itself: It is true that we, in a sense, discover patterns in that we see what is already there!

However, to capture these patterns and describe them in a good way, we need to do some patterns design work—invent if you will.

Pattern specificity and validity

How specific should the patterns be and how do we determine the validity of a pattern?

Patterns that are too generic are of no practical use, and the ones that are too specific (to a domain) certainly have difficulties in describing generically good solutions or providing generically useful information. The distinction between generic and specific patterns represents a delicate balancing act. In fact, this discussion took up a large portion of the workshop, evolving into related questions about pattern validity.

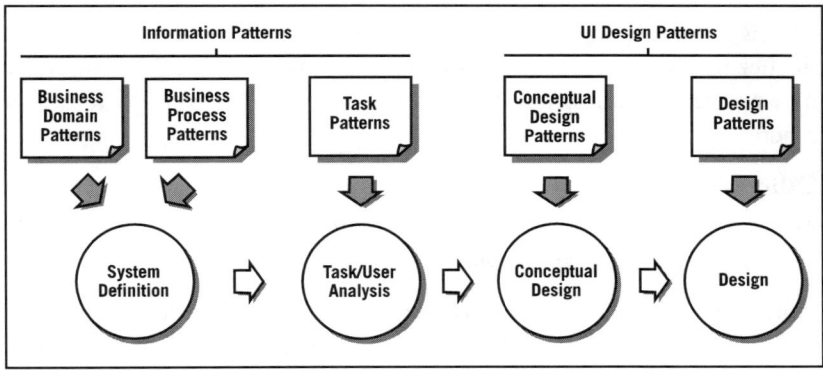

Fig. 2: The two parts of a task pattern.

Let us say, for instance, that a specific task pattern includes a number of influencing factors and points to a conceptual design pattern. Is the task pattern in itself useful, and will the conceptual design it suggests be good if some—but not all—factors apply?

We discussed this issue at length and came up with an idea that seems to make sense: Let's suppose that task patterns are somehow divided in two parts, one that is generic and points to one or more conceptual design patterns and one that is domain-specific.

This means that we can always be sure that the forces that are listed in the generic part (framing forces) apply to the suggested conceptual design patterns. PSA talks of conceptual designs that are data- or task-oriented (a specific system may well be designed for either orientation), and it was agreed that it is relevant to also add an agent (actor)-oriented view. Taking this into consideration, the general opinion among the participants was that the number of basic conceptual designs was still limited, and that a conceptual design will also imply the navigation system.

The domain-specific part could contain more information that might or might not apply, and also suggest a number of resulting questions, emerging from the forces (differentiating forces). Examples should also be provided, if possible, in order to promote practical use.

The suggested division must support that need for domain-specific information to somehow carry to the conceptual design and the screen design and not be lost on the way.

This is an interesting idea that would solve some problems and will be further investigated. It must, however, be possible to do this smoothly, in a way that ensures that we keep patterns intuitive and makes them practically usable.

Other questions

Do we stand to lose the inner quality aspect of patterns by a process approach and do we expect too much from patterns?

During the day, a question recurred asking if the process approach would make us lose the very idea of patterns by introducing a mechanistic perspective on patterns and their application. That is definitely not the idea behind PSA, and our discussions made clear that it is important to point this out. We're suggesting pattern support for the process. To benefit by it, one should not have to follow any specific process step-by-step. The patterns are just one source of information—providing what they are intended to provide—invariant information for good design. Also, if the patterns get too complicated and difficult to use, we certainly have not done a good job.

Do we need a meta language for pattern usage?

The idea of describing good use of patterns by using some kind of meta pattern language is an interesting thought. It could be used for describing how the information contained in the patterns is best put to use—how analysis is performed and how differences are handled. The question is interesting enough to look into further; however—and this is very important—we must still make sure that we do not overcomplicate things or lose the very idea behind patterns.

We have the feeling that we could very easily end up losing the basic concept if we increase the complexity of pattern support and imply a strict procedure for their use.

Special thanks

We would like to thank the participants: Kay Aubrey, George Casaday, Tom Dayton, Michael Muller, Jenifer Tidwell, Peter Warren, and Daniel Wildman for their great participation and sharing their thoughts on patterns.

Special thanks to Michael Muller and JoAnn Hackos for their comments on the preliminary version of this workshop report.

Also, a very special thanks to Tom Erickson for his comments on the first version of our pattern-supported approach.

Up-to-date comment on this workshop report

We received very valuable input from the workshop described in this report, and over the months since, also from various discussions with people from the usability and patterns communities, from practical application and from giving tutorials.

As a result of this, the PSA has evolved and we would like to take the opportunity to briefly comment on this, tying back to some of the workshop discussions.

For detailed information about updates, please refer to our Web site: http://www.gespro.com/lafrenid/patterns.html

So what has happened?

The framework as described in the report is still unchanged and the basic ideas and concepts seem robust. In the following, we will describe some changes that have been done, or are in the progress of being introduced.

Task patterns started out as information patterns (not providing a solution), but now include an interaction design solution. This turned out to be a natural step, since there are a number of interaction design decisions made based on the task, and task patterns was the place to provide this knowledge.

We have not focused our recent work on the other information patterns, but it might well be the case that there are corresponding design solutions for them as well, which would mean abandoning the pure information patterns as defined previously.

However, the information patterns all contribute to the analysis of the current situation, which still puts them in a class by themselves, regardless of if they also provide a solution on some level or not. The workshop conclusions still hold, and the distinction is still there— though maybe not as rigid as before.

We also decided to introduce subtask patterns. These describe the smaller tasks that are part of the complex task described in the task pattern.

The reasons for this are mainly:

- Complexity. Any complex task (which is what we have been working with) is in fact built up by a number of smaller tasks that may be quite diverse, and it's impossible to effectively describe all parts and the whole in one pattern. Subtask patterns are generic, making up building blocks for more complex tasks, while at the same time having their own pattern description with forces, related activity patterns and structure and navigation design patterns (see below).

- Structure and navigation . . . which brings us to the change we have made regarding conceptual design patterns. During the workshop, we discussed the value of conceptual design patterns and also the validity, specifically regarding determining the validity of a conceptual design pattern if not all parts of the task pattern apply.

We still want to provide design knowledge regarding the organization of information and task, but we abandoned the concept of conceptual design patterns. The reason for this is that in practicality, they turned out to be too abstract to be really useful. Instead we have taken a different approach to the same problem. We substituted the conceptual design patterns with what we call structure and navigation patterns. We think there is a strong correlation between how you navigate in a system, and how the information is structured, and we chose to concentrate on determining a suitable navigation model (for a task, there is generally one or more suitable navigational styles) and let this imply the structure of information and functions. This seems to be a more straightforward and intuitive way of solving the structure design problem.

Also the previously described division between tasks and subtasks allows us to suggest an overall navigational/structural approach (for the task) including other navigation/structure models (for included subtasks) which addresses the problem with different levels of detail and underoptimization regarding structure, as discussed in the workshop report.

Regarding validity, we have in some way pursued the workshop idea of a generic part in the task pattern, for which we can say that pointers are valid. Instead of having an explicit part of the pattern that is generic, we instead let the domain-specific patterns be derived from generic patterns. (Example: the patterns Patient's Record Review and Criminal Record Review may both be derived from the generic pattern Case Folder Analysis, but the domain-specific information will hopefully differ. . . .) These generic patterns can certainly by themselves provide information, but they are explicitly stripped of all domain-specific information, and will thereby be too abstract to be put into practical use (for analysis of user, task and context) within a specific domain.

In the end, we think that the patterns format in itself (providing rationale) will promote intelligent usage—we do not want to write a cookbook!

References

Appleton, B. "Patterns and Software: Essential Concepts and Terminology". http//www.enteract.com/~bradapp/docs/patterns-intro.html.

Erikson, T. "The Interaction Design Patterns Page." www.pliant.org/personal/tom_erikson//interaction patterns.html.

Erickson, T. (1997). "Putting It All Together: Pattern Languages for Interaction Design." Workshop CHI 97 Conference, Atlanta, GA, 1997.

Fowler, M. (1997). *Analysis Patterns: Reusable Object Models.* Addison Wesley, Menlo Park, CA.

Granlund, Å. and Lafreniere, D. (1999). "The PSA Approach." http//www.gespro.com/lafrenid/patterns.html.

Lafreniere, D (1998). "From Entity-Relationship Diagram and Task Analysis Diagram to User-Interface Architecture. Workshop presentation. UPA 98, Washington, D.C.

Lea, D. (1998). "Patterns Discussion FAQ". g.oswego.edu/dl/pcl-FAQ/pd-faq.html.

Louerio, K. and Plummer, D. (1999). *AD Patterns: Beyond Objects and Components.* Research Note #COM-08.0111. Gartner Group.

Tidwell, J. "Common Ground: A Pattern Language for Human-Computer Interface Design." http//www.mit.edu/~jtidwell/interaction-patterns.html.

Web Site Evaluation

Following a Fast-Moving Target: Recording User Behavior in Web Usability Testing

Laurie Kantner,
Tec-Ed, Inc.

Web usability testing presents special challenges for taking notes about user behavior. Capturing user behavior accurately and completely for immediate reporting is difficult to accomplish in "real time" with fast-clicking and complicated user interface elements to track. However, it is critical for the rapid usability feedback that Web site developers demand. This article describes the challenges of capturing user-Web-site behavior and the recording methods that usability specialists at Tec-Ed have developed.

Context for Web usability testing

The development cycle for many Web sites is more incremental than for other types of software. A new version of a Web site does not require announcing or shipping—users simply encounter the new version when they next visit the site.

Web site owners are concerned when server logs show that pages they want users to see are receiving very few hits. Knowing how users navigate the site and where they get sidetracked are vital to paving a smoother path to the key pages.

However, the ability of an organization to make incremental improvements to Web sites makes developers unwilling to wait long for usability feedback. At many organizations, the usability specialists perform "off the cuff" expert evaluation to meet the demands for instant feedback.

Web sites designed to generate revenue generally have more structured development processes. Usability improvements to such sites have more visible impact on the bottom line—for example, increasing the number of software downloads from a site. Thus, revenue-generating sites are likelier candidates for laboratory usability testing than information-only sites. Web site use is difficult to record.

In most Web sites, as in many hypertext systems, users have enormous freedom of action. Clicks can happen quickly, and we cannot always tell at the moment if the user's action is germane to the issues of concern. Thus, we must err on the side of recording too much, not too little. In essence, we must record every click so that we can retrace the user's steps. The difficulties of recording this behavior are:

- Users of Web sites can take numerous paths to reach their goal. It is difficult to develop shorthand for identifying so many paths.

- Links and buttons that have similar names but different destinations are rampant on the Web. We must record these user choices accurately for later analysis, yet, a shorthand for identifying links and buttons is difficult with so many similar-sounding names.

- Users can traverse many individual Web pages to reach their destinations, and recording these locations is important for determining where a problem exists. However, recording Web page titles is difficult with pages that lack titles or have wordy or awkward titles.

- Web page users often cycle through pages repeatedly, trying to get to their desired destinations. Recording return visits is not only important for identifying where users are getting lost or confused, but also difficult because users tend to speed up when they repeat steps they've already taken.

- Recording detailed behavior on dynamically generated pages is a challenge in real time, especially for unanticipated pages created "on the fly." An accurate recording of events leading to display of that page is crucial to replicating the user's behavior should the observer's notes about that page lack sufficient detail.

Whether a usability specialist takes notes on the spot or from viewing videotapes or Lotus Screencam recordings later, it is physically difficult to label every choice the user makes. Tec-Ed usability specialists prefer to take detailed notes in real time to have ready data for quick analysis and reporting. Taking notes at the user's pace, however, further aggravates the problem of describing the user's choices precisely—distinguishing identically named buttons and links while the user has moved to another page is a problem we encounter all too frequently.

Automated data-collection methods can delay results reporting

Automated methods for recording user behavior are popular with today's usability professionals. These methods include videotaping, Lotus Screencam recording, data-logging software, and server log files. All these methods reduce how much note-taking the usability specialist must perform during the session. However, they do not meet all the criteria for collecting accurate data that will be available for immediate analysis and reporting:

- **Videotaping.** The advantage of videotaping, for both Web-based and non-Web-based software, is that it records fast-paced user activity that handwritten notes might miss. It also captures user commentary and cursor-pointing behavior. Thus it promotes accuracy and completeness. However, a single fixed video camera may not capture all user behavior. Most important, this method delays results reporting because the usability specialist must review tapes to formulate a complete set of data to analyze. Depending on whether you take notes from the videotape or actually transcribe it, compiling data from videotapes can take from one to eight hours per hour of tape.

- **Screencam recording.** As with videotaping, this method records fast-paced user activity accurately and completely. While it captures cursor-pointing behavior, it does not capture user commentary. Thus it is slightly less complete than videotaping. Reviewing recordings also delays reporting of results.

- **Data-logging software**. The advantages of using data-logging software are that the note-taker can record textual notes more quickly than writing by hand, and can code observations into categories such as "Error" or "Observer Comment." These advantages mean the data is more complete and already somewhat sortable into categories for more immediate analysis. This method still does not solve the problem of labeling user choices at the user's pace.

- **Server log files.** This automated method is unique to Web software, and its advantage is it records a lot of detail—so much so that one might think every keystroke is captured. However, log files in fact miss important information: client-side events such as pages displayed from cache (return visits), cursor-pointing behavior, and JavaScript activity. Equally disadvantageous are the large amounts of time required to synthesize the individual records into episodes.

Until automated tools are developed that can convert videotaped sessions or electronically captured user keystroke sequences into organized tabulations of task episodes—complete with timing information and user commentary—additional note-taking will be required, and time will be needed to convert the recorded material into a database of observations for analysis. For now, we seek the simplest method that meets our requirements.

Writing by hand is not exciting in the world of technology. Yet Tec-Ed specialists have found time-savings from taking detailed notes during the actual sessions, and have developed session note-taking methods to meet the special challenges of Web usability tests.

A note-taking method that works

For lab test observation, Tec-Ed has migrated to a hybrid method of note-taking that meets our requirements for accurate, complete information, as well as ready data for immediate analysis and reporting. We have adapted this method to meet the special challenges of Web usability testing.

In usability tests where the user's task is linear or contains few branches, we can record user behavior using a checkoff data-collection form. This type of data-collection device requires us to consider the range of outcomes that may occur, based on our use of the software and the development team's concerns, and pre-organize these outcomes into multiple-choice lists on which the test administrator or observer can check off items as they occur. These data-collection forms also provide white space for jotting down user comments or "custom" observations.

The success of this data-collection method depends on six factors:

~ The usability test is of a defined set of Web pages; that is, tasks do not include learning the site for unknown destinations.

~ A usability colleague reviews the data-collection form to identify additional choices or improvement in presentation of choices for fast recording.

- The dry-run of the usability test includes a dry-run of the data-collection form.
- The usability specialist practices using the data-collection form so that finding and checking off items does not mean missing user actions. The usability specialist must become skilled at handling these forms unobtrusively. We use this method both in the room with the user and in a separate observation room. Fortunately, our experience is that the user's attention remains focused on the screen, not on our note-taking.
- The testing or observation room has adequate surface area for handling the paper.
- The usability specialist is skilled in writing quickly, because ultimately, the data-collection form cannot anticipate every possible behavior of the user's chosen chronology.

Of course, the data-collection form also includes prompts for Start Time and End Time, and the usability specialist has practiced filling these in diligently.

Examples of data-collection forms for software testing

Here is an example of part of a filled-in user-data-collection form for software in which the user's task is extremely linear:

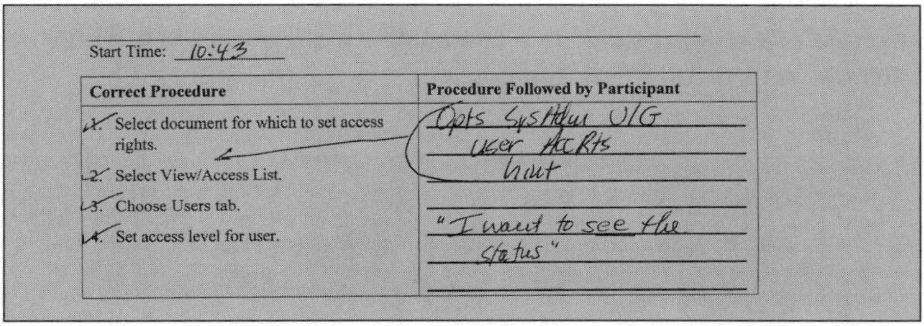

Start Time: 10:43

Correct Procedure	Procedure Followed by Participant
1. Select document for which to set access rights.	Opts SysAdmin U/G User AccRts hunt
2. Select View/Access List.	
3. Choose Users tab.	
4. Set access level for user.	"I want to see the status"

Figure 1. Filled-in data-collection form for a linear task.

In the above example, the usability specialist would either place a checkmark next to the steps completed or write additional steps on the right side, drawing an arrow to the left side where they occurred within the sequence of correct steps.

Here is an example of part of a data-collection form for software in which the user can make more branching choices:

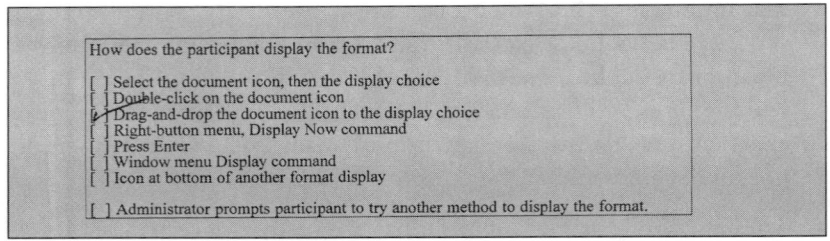

Figure 2. Filled-in data-collection form for a branching task.

In this example, the usability specialist marks the appropriate checkbox. For lists where users might perform multiple items, the usability specialist numbers the items in chronological order. The space at the right is available for recording additional information such as user comments. The advantage of the checkoff data-collection form is that it enables the usability specialist to capture predictable actions quickly and fairly accurately. Most important, after the session the results are easy to tally for analysis and reporting. The disadvantage of this form is that users can make surprising choices, and then the usability specialist has to annotate the form. However, those instances themselves are extremely informative in telling us where everyone's expectations were incorrect.

As mentioned earlier, we add the space to check off forms for additional observations or capturing user comments, with labels identifying the intended information. For example:

Figure 3. Additional comments on a check-off form.

Another disadvantage of this form style is that these forms can occupy lots of pages; skill and practice are required in handling them with ease.

Some usability specialists prefer to use blank paper and take "stream of consciousness" notes or to use a computer and simply type into a blank document. These methods can generate complete notes but usually offer no pre-organization of information from which to tally quickly.

Adapting the data-collection form for Web studies

Despite our efforts to create a usable checkoff data-collection form, it has proved insufficient for note-taking of participant behavior during Web usability tests. Why?

- To provide a list containing every possible choice a user can make on a Web page would require absurdly long lists of items—and the observer couldn't possibly find the desired items quickly enough. Our data-collection forms already occupy 10 to 15 pages for a 2-hour session; we believe that is enough pages to master.

- Web pages continue changing right up to the test day. However carefully we label the Web-page choices we want to track, some of those choices will change or even go away on testing day.

Therefore, we replaced the detailed lists of checkbox items with printouts of the Web pages on which we record user actions. These printouts can be generated at the last minute to reflect the latest updates. We photo-reduce the Web page printouts to create a column of blank space at the right. For the first visit, we place a "1" next to the first link chosen, then put the number "2" on the page this link goes to. For a repeat visit, we write user actions in the blank space, pointing to links for clarity. We also use the blank space to record user comments (see figure 4).

To facilitate page-flipping, the usability specialist is wise to use sticky notes on the edges of key pages. All possible pages of interest must be included in the packet.

In addition to Web page printouts, we pre-organize some of the expected outcomes into higher-level data-collection forms with spaces for tabulations, to record more summary-level observations. After a user completes a task or subtask, we use the data-collection forms to summarize what happened for quick post-test reporting.

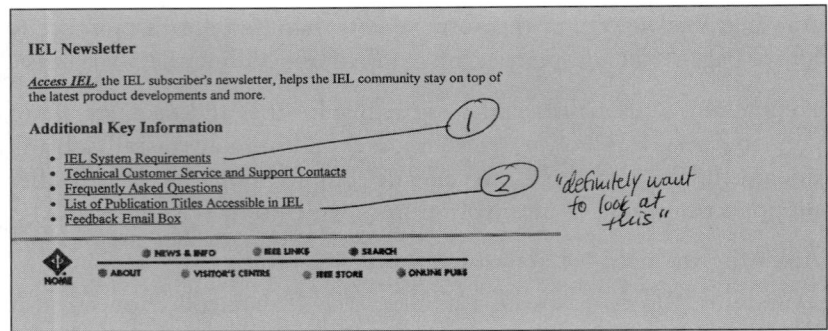

Figure 4. Blank spaces for user comments.

Did the participant recognize "[Product Name]" as what s/he is looking for? ✓Yes __No

Did the participant want to read [Product] background information on-line? ✓Yes __No

Did the participant read the URLs? __Yes ✓No

Did the participant ever see the [Product] home page? __Yes ✓No
Recognize it? __Yes __No

Figure 5. Pre-organized outcomes on a data-collection form.

Figure 5 is an example of pre-organizing outcomes on a data-collection form filled out on completion of a session.

In the above case, the data-collection form was for a study in which the user was to find and download a type of software without knowing its exact name. The user could potentially visit up to 70 Web pages to complete the test tasks. Our goal was to record all user choices, while also making sure we captured the answers to the above questions that addressed key concerns.

As test administrator in the same room with users, I kept one stack of Web page printouts and one set of checkoff data-collection forms sitting at a working surface that enabled me to alternate between the two sets as needed. The user was focused on the screen and rarely looked at my recording or me. If I got behind, I caught up during the time between tasks when the participant was filling in a questionnaire.

Analyzing the notes after the session

At the end of the first session, the usability specialist can begin designing intermediate tabulation forms for collecting all of the data in one place for analysis. The checkoff forms simplify choice and timing tabulations.

The most time-consuming part is going through the annotated Web page printouts and recording events considered significant. However, using annotated printouts to accomplish this task is far less time-consuming than reviewing videotapes.

Figure 6 is an example of an intermediate tabulation of notes taken for a Web page study, using the method described above. (Note that we occasionally also spot-viewed videotapes to confirm or fill in our notes; this spot-checking took less time than reviewing the tapes to create notes.)

Recognizes "[Product Name]"?	Participant	First Link Chosen from Home Page	Number of Pages Visited
Pilot	LinkA	10	Already knew its name
1	LinkB	8	Figured it out
2	LinkA	8 (hint)	Figured it out (hint)
3	LinkA	12 (hint)	Figured it out (2 hints)
4	LinkA	11 (gave up)	Had to be told directly
5	LinkC	10 (hint)	Figured it out (hint)
6	LinkD	3	Figured it out
7	Search	9 (hint)	Had to be told directly
8	LinkA	13+	Used search
9	LinkE	10 (hint)	Figured it out (3 hints)
10	LinkB	9 (hint)	Figured it out (hint)
11	LinkA	7	Figured it out
12	LinkA	8	Already knew its name

Figure 6. Intermediate tabulation of notes.

In this example, we see that LinkA was a popular but less productive choice than LinkD, and that many users required hints to succeed. The data for the First Link Chosen from Home Page column and the Number of Pages Visited column came from analysis of Web-page-printout annotations, and the data from the Recognizes [Product Name] column came from the checkoff data-collection forms.

Conclusion

Web site usability testing requires fast, accurate data-collection in an accessible format if the usability team wants to deliver immediate results—which Web site developers demand. Using a combination of Web-page printouts for detailed data-collection and checkoff data-collection forms for pre-organizing higher-level outcomes gives the usability specialist the range of instruments needed to give results fast.

The Web page printouts enable the usability specialist to track detailed Web usage data in "real time" without the added difficulty of labeling the user's choices precisely. The data-collection forms answer higher-level concerns, summarizing the bottom line of how the user worked with the site.

This method of collecting data about Web usage works for all sizes of Web sites, although extremely large sites increase the difficulty of making sure all pages the user might visit are represented in the data-collection forms. This method also accommodates those last-minute changes to Web sites we have learned to expect. We still ask the Web site developers to keep the site unchanged during the actual days we conduct the usability testing. And sometimes it happens!

The IEEE's IEL Web page is shown with permission from IEEE.

Assessing Web Site Usability from Server Log Files

Laurie Kantner,
Tec-Ed, Inc.

Web log file analysis began as a way for IT administrators to ensure adequate bandwidth and server capacity on their organizations' Web sites (Wilson). Log file analysis has advanced considerably in the past five years, with companies now mining log files for finer-grained detail about visitor profiles and buying activity. Organizations are now seeking ways to use log files to learn about the usability of their Web sites—that is, how successfully visitors meet their specific information or transaction goals there.

Log file data can offer valuable insight into Web site usage. It reflects actual usage in natural working conditions compared to the artificial setting of a usability lab. It represents the activity of many users, over a potentially long period of time, compared to a limited number of users for an hour or two each.

Although these advantages are an overpowering reason to investigate log file data for usability purposes, the data as it is often collected today in fact answers very few usability questions. Log file analysis is best used in a well-structured program of continuing usability research to discover data that complements—or spurs—studies of greater depth using other usability methods.

This paper explores the limitations of log file data for usability analysis. It briefly describes server log file analysis, discusses the requirements for log file data to yield usability data, and presents ways to integrate log file analysis into the usability engineer's tool kit. The paper builds on Tec-Ed's experience in performing log file analysis and other usability research for our clients as well as our own Web site.

What is server log file analysis?

Server log files are records of Web server activity. They provide details about file requests to a Web server and the server response to those requests. In the access log, which is the main log file, each line describes the source of a request, the file requested, the date and time of the request, the content type and length of the transferred file, and other data such as errors and the identity of referring pages.

Here is a portion of the log file for Tec-Ed's Web site, showing one home page access. The column headings identify the types of

```
Source of Request (Host) Date and Time of Request  Page Requested (HTTP protocol) Status Code Number of Bytes
Referring Page    Browser    Platform
pm471-46.dialip.mich.net - - [24/Oct/1999:19:13:44 -0400] "GET /images/tagline.gif HTTP/1.0" 200 1449
"http://www.teced.com/" "Mozilla/4.51 [en] (Win98; I)"
pm471-46.dialip.mich.net - - [24/Oct/1999:19:13:44 -0400] "GET /images/bkgrnd.jpg HTTP/1.0" 200 10659
"http://www.teced.com/" "Mozilla/4.51 [en] (Win98; I)"
pm471-46.dialip.mich.net - - [24/Oct/1999:19:13:44 -0400] "GET /images/yellow_bit.gif HTTP/1.0" 200 280
"http://www.teced.com/" "Mozilla/4.51 [en] (Win98; I)"
pm471-46.dialip.mich.net - - [24/Oct/1999:19:13:44 -0400] "GET /images/TE_logo.gif HTTP/1.0" 200 1292
"http://www.teced.com/" "Mozilla/4.51 [en] (Win98; I)"
pm471-46.dialip.mich.net - - [24/Oct/1999:19:13:52 -0400] "GET /images/site_map.gif HTTP/1.0" 200 714
"http://www.teced.com/" "Mozilla/4.51 [en] (Win98; I)"
pm471-46.dialip.mich.net - - [24/Oct/1999:19:13:53 -0400] "GET /images/home_00.gif HTTP/1.0" 200 43
"http://www.teced.com/" "Mozilla/4.51 [en] (Win98; I)"
pm471-46.dialip.mich.net - - [24/Oct/1999:19:13:53 -0400] "GET /images/home_00.gif HTTP/1.0" 200 43
"http://www.teced.com/" "Mozilla/4.51 [en] (Win98; I)"
pm471-46.dialip.mich.net - - [24/Oct/1999:19:13:55 -0400] "GET /images/use_eval_hbut.gif HTTP/1.0" 200
747 "http://www.teced.com/" "Mozilla/4.51 [en] (Win98; I)"
pm471-46.dialip.mich.net - - [24/Oct/1999:19:13:55 -0400] "GET /images/marcom_hbut.gif HTTP/1.0" 200
911 "http://www.teced.com/" "Mozilla/4.51 [en] (Win98; I)"
pm471-46.dialip.mich.net - - [24/Oct/1999:19:13:55 -0400] "GET /images/contact_us.gif HTTP/1.0" 200 659
"http://www.teced.com/" "Mozilla/4.51 [en] (Win98; I)"
pm471-46.dialip.mich.net - - [24/Oct/1999:19:13:56 -0400] "GET /images/uid_hbut.gif HTTP/1.0" 200 637
"http://www.teced.com/" "Mozilla/4.51 [en] (Win98; I)"
pm471-46.dialip.mich.net - - [24/Oct/1999:19:13:56 -0400] "GET /images/c+p_hbut.gif HTTP/1.0" 200 699
"http://www.teced.com/" "Mozilla/4.51 [en] (Win98; I)"
pm471-46.dialip.mich.net - - [24/Oct/1999:19:13:57 -0400] "GET /images/who_is_hbut.gif HTTP/1.0" 200
619 "http://www.teced.com/" "Mozilla/4.51 [en] (Win98; I)"
pm471-46.dialip.mich.net - - [24/Oct/1999:19:13:57 -0400] "GET /images/whats_new.gif HTTP/1.0" 200 375
"http://www.teced.com/" "Mozilla/4.51 [en] (Win98; I)"
pm471-46.dialip.mich.net - - [24/Oct/1999:19:13:58 -0400] "GET /images/doc_hbut.gif HTTP/1.0" 200 1015
"http://www.teced.com/" "Mozilla/4.51 [en] (Win98; I)"
pm471-46.dialip.mich.net - - [24/Oct/1999:19:14:30 -0400] "GET /Octmochi.htm HTTP/1.0" 200 7207
"http://www.teced.com/" "Mozilla/4.51 [en] (Win98; I)"
pm471-46.dialip.mich.net - - [24/Oct/1999:19:14:32 -0400] "GET /images/mocslid.gif HTTP/1.0" 200 1407
"http://www.teced.com/Octmochi.htm" "Mozilla/4.51 [en] (Win98; I)"
pm471-46.dialip.mich.net - - [24/Oct/1999:19:14:35 -0400] "GET /images/getacro.gif HTTP/1.0" 200 712
"http://www.teced.com/Octmochi.htm" "Mozilla/4.51 [en] (Win98; I)"
pm471-46.dialip.mich.net - - [24/Oct/1999:19:15:03 -0400] "GET /PDFs/mochi99.pdf HTTP/1.0" 200 64667
"http://www.teced.com/Octmochi.htm" "Mozilla/4.51 [en] (Win98; I)"
pm471-46.dialip.mich.net - - [24/Oct/1999:19:16:39 -0400] "GET /PDFs/mochi99.pdf HTTP/1.0" 200 64667
"http://www.teced.com/Octmochi.htm" "Mozilla/4.51 [en] (Win98; I)"
pm638-17.dialip.mich.net - - [24/Oct/1999:19:52:23 -0400] "GET /PDFs/mochi99.pdf HTTP/1.0" 200 64667
"http://www.teced.com/Octmochi.htm" "Mozilla/4.51 [en] (Win98; I)"
```

Figure 1: Log file sequence for Tec-Ed's Web site, showing initial home page access.

information recorded in this log file. Two problems make log file analysis for usability assessment difficult. The first is insufficient data in the log file; the second is extraneous data in the log file. You'll need to work with your organization's technical staff to make sure the data you want to analyze is logged and summarized (see "Log File Data Required to Yield Usability Data," next).

More and more log file analysis and reporting tools are becoming available. Although the majority of new tools focus on marketing research requirements, some of these tools produce databases that can be helpful in usability analyses. If your log files are small, you can even work directly with raw log file data using a tool such as a spreadsheet program. For example, Tec-Ed's log file for October 1999 is composed of 14,000 individual entries. This log file is considered small by industry standards.

Log file data required to yield usability data

The data collected in log files can vary from one server to another. To best complement the wide variety of analyses desirable in a usability engineer's tool kit, the log file should be comprehensive and transparent. Standard log file formats fall short of this goal; you'll need to work with your organization's technical staff and the site developers to define what data goes into the log file.

The ideal log file for usability analysis

The ideal log file for usability analysis contains data you can use to learn:

- Who is visiting your site. You want unique visitor identification so you know whether a visitor is returning to your site.

- The path visitors take through your pages. With knowledge of each page a visitor viewed and the order, you can identify trends in how visitors navigate through your pages. You also want to know what element (link, icon) a visitor clicked on each page to go to the next page.

- How much time visitors spend on each page. A pattern of lengthy viewing time on a page might lead you to deduce the page is very interesting—or very confusing.

- Where visitors are leaving your site. The last page a visitor viewed before leaving your site might be a logical place to end the visit, or it might be a place where the visitor bailed out.

- ~ The success of users' experiences at your site. Purchases transacted, downloads completed, and information viewed are concrete indicators of tasks accomplished.

In other words, you want enough data to reconstruct the entire "episode" of the user's visit to your site. (Often the term "session" is used in log file analysis tools; however, a session might be a partial episode for reasons discussed in "Episodes vs. Sessions" later in this paper.) Unfortunately, the information you want to know might or might not be in the log file. Even if it is in the log file, other data might make it difficult to interpret.

Why log files commonly fall short

Log files were designed to produce site-level performance statistics. It's thus no surprise they can't provide even the minimum information needed to effectively investigate a potential usability problem. Here are some specific ways log files provide insufficient or misleading data:

- ~ Who is visiting your site. For you to know who is visiting your site, the log file must contain a person ID such as a login to the server or to the user's own computer. However, most Web sites do not require users to log in, and most Web servers do not make a "back door" request to learn the user's login identity on his/her own computer.

  The log file does provide information about the requesting host. This information might identify a single-user computer, enabling unique identification for episode tracking. More often, it is an IP address temporarily assigned by an Internet service provider (ISP) or corporate proxy server to a user's TCP/IP connection to your site, preventing unique identification. The information also might be an address for a shared computer or for a shared security gateway.

- ~ The path visitors take through your pages. The path that visitors follow within your site is clear if the log file contains an entry for every page viewed. However, when browsers are set to view pages from cache (usually the default), or when corporate or ISP servers retrieve pages from a central cache, then some pages will not be logged by the Web server and the log file will have gaps. For example, with caching, pages viewed using the Back button typically are not logged.

What's more, it's important to identify the link used to move to another page when two or more choices are provided on a page.

Having each link point to a different but synonymous page name, such as page.htm and page.htm/, enables identification of which link was chosen. (It can also lead to site maintenance difficulties if two pages are provided for the same content.) If one of the links is an image map, the log file might contain the screen coordinates for the click, enabling reconstruction of this information. Without these techniques, one can only guess which link the user clicked to reach another page.

In addition, nothing appears in the log file when visitors arrived at a page by typing its URL, using a bookmark, or following an e-mail link (Drott). In these cases one can try to infer from Referrer data.

~ How much time visitors spend on each page. The log file records the time when a data transmission was initiated, but not the time when the transfer was completed. In addition, it is unclear when during the download process the user began viewing a page. However, by comparing the timestamps of the current request and the next request, you can calculate roughly how much time a visitor is spending on a page—unless the visitor walks away while the computer is displaying the page. Some timing details may also be obtained by analyzing the transmission of graphics files associated with a page.

~ Where visitors are leaving your site. The log file records the last page transferred by the server for that user session, but there are two reasons why it might not be the last page viewed. First, the last page viewed may have been displayed from cache. Second, the user may have left his/her workstation for a period of time that exceeds what the log analysis software regards as a session.

~ The success of users' experiences at your site. Alas, the ultimate usability question—how successful was the user experience at your site?—cannot usually be answered by log file statistics alone. If the question is equivalent to, "Did the user complete a purchase transaction?" or "Did the user successfully download a file?" the answer is easier to deduce. However, if you're asking, "Did the user find the information he/she needed?" the answer requires additional research that can be informed by log file data.

Here is another portion of the log file for Tec-Ed's Web site, annotated with questions raised by the log file data.

This address belongs to a block of IP addresses. We do not know who the user is nor do we know if the user revisits the site.

36 minutes elapsed—was the previous page interesting? Confusing? Or did the user leave the computer?

The sequence starting here reflects gaps in the page sequence (the referring page is not the previous page viewed). We assume use of the Back key.

53 minutes on the Site Map page. Again, did the user leave the computer?

Source of Request (Host) Date and Time of Request Page Requested (HTTP Protocol)   Referring Page   Browser   Platform
198.108.69.19 - - [29/Oct/1999:08:07:39 -0400] "GET / HTTP/1.0" 200 6297 "-" "Mozilla/4.0 (compatible; MSIE 5.0; Windows 98)"

.(downloading of gifs removed for readability throughout this episode)

198.108.69.19 - - [29/Oct/1999:08:08:58 -0400] "GET /c_and_p.html HTTP/1.0" 200 10754 "http://www.teced.com/" "Mozilla/4.0 (compatible; MSIE 5.0; Windows 98)"
198.108.69.19 - - [29/Oct/1999:08:10:13 -0400] "GET /index.html HTTP/1.0" 200 6297 "http://www.teced.com/" "Mozilla/4.0 (compatible; MSIE 5.0; Windows 98)"

.(requests from a different IP address)

198.108.69.19 - - [29/Oct/1999:08:46:13 -0400] "GET /whois.html HTTP/1.0" 200 5137 "http://www.teced.com/" "Mozilla/4.0 (compatible; MSIE 5.0; Windows 98)"
198.108.69.19 - - [29/Oct/1999:08:48:02 -0400] "GET /use_eval.html HTTP/1.0" 200 6209 "http://www.teced.com/" "Mozilla/4.0 (compatible; MSIE 5.0; Windows 98)"
198.108.69.19 - - [29/Oct/1999:08:48:57 -0400] "GET /uid.html HTTP/1.0" 200 4114 "http://www.teced.com/" "Mozilla/4.0 (compatible; MSIE 5.0; Windows 98)"
198.108.69.19 - - [29/Oct/1999:08:49:58 -0400] "GET /ue-ust.html HTTP/1.0" 200 7272 "http://www.teced.com/use_eval.html" "Mozilla/4.0 (compatible; MSIE 5.0; Windows 98)"
198.108.69.19 - - [29/Oct/1999:08:50:58 -0400] "GET /ue-fs.html HTTP/1.0" 200 6274 "http://www.teced.com/ue-ust.html" "Mozilla/4.0 (compatible; MSIE 5.0; Windows 98)"
198.108.69.19 - - [29/Oct/1999:08:53:18 -0400] "GET /sitemap.html HTTP/1.0" 200 5253 "http://www.teced.com/uid.html" "Mozilla/4.0 (compatible; MSIE 5.0; Windows 98)"

.(requests from a different IP address)

198.108.69.19 - - [29/Oct/1999:09:46:39 -0400] "GET /doc_help.html HTTP/1.0" 200 5966 "http://www.teced.com/" "Mozilla/4.0 (compatible; MSIE 5.0; Windows 98)"
198.108.69.19 - - [29/Oct/1999:09:46:43 -0400] "GET /images/document_help.gif HTTP/1.0" 200 2620 "http://www.teced.com/doc_help.html" "Mozilla/4.0 (compatible; MSIE 5.0; Windows 98)"
198.108.69.19 - - [29/Oct/1999:09:47:23 -0400] "GET /mar_com.html HTTP/1.0" 200 6423 "http://www.teced.com/doc_help.html" "Mozilla/4.0 (compatible; MSIE 5.0; Windows 98)"
198.108.69.19 - - [29/Oct/1999:09:47:24 -0400] "GET /images/market_comm.gif HTTP/1.0" 200 2374 "http://www.teced.com/mar_com.html" "Mozilla/4.0 (compatible; MSIE 5.0; Windows 98)"

Figure 2: Log file sequence showing time and sequence gaps in page requests.

The rest of this paper explores ways to capture visitor, path, time, and user-success data for assessing Web site usability, either through log file analysis or with the help of log file analysis.

Visitor identification data

Obtaining user identity information—enough to know one user from another, not to intrude on a user's privacy—enables usability engineers to distinguish data representing regular users from data representing new users or infrequent users. Distinctions among users according to frequency of use are valuable in improving Web site usability—the problems these audiences have can be different, and a solution for one audience might create problems for another audience.

Two methods are available for obtaining a user's identity for log file purposes: user registration/login and cookie files. Asking users to identify themselves, either explicitly through login screens or implicitly through accepting cookies, risks users becoming unhappy with a Web site. Your organization must weigh the trade-offs and reach a happy medium between your ability to track users and their happiness.

User registration/login

When a user supplies a login identity to a Web server, that information is stored in each log file record for that user's subsequent activity at the site. This information enables tracking of the user's Web site experience.

Most users say they dislike logging in to a site, and many report forgetting their passwords or supplying "bogus" identities. However, in many usability studies conducted by Tec-Ed of transaction-oriented Web sites, study participants say they understand the business reasons behind why a site asks for identification.

There are two reasons to ask earlier rather than later in a session for the user's identity:

- For the usability analyst: to attach the user's identity to as many actions at the site as possible.

- For the user: to avoid the "surprise factor." Users express annoyance at progressing through many pages of a site, only to arrive at a registration or login page when they least expect it.

Users do not expect to provide identity information at informational sites—after all, anonymous lurking is the legacy of the Web. In these cases a different means is available to identify visitors, which is also commonly used at transaction-oriented sites: cookie files.

Cookie files

In many environments, the Web server can record information transferred to and from "cookie" files into the log file. This information often includes some kind of user identity information that the server has passed to the cookie file, as well as information about transactions. (The server reads this information upon the user's next action or next access to the site.) Having this information in the log file enables improved tracking of individual user sessions and analysis of regular versus infrequent users. However, there are still difficulties with shared computers configured so that all users appear as a single user, individuals with multiple computers, and users who periodically clean up their cookie files or who install new software.

Here is a sample scenario of cookie file detection.

Cookie File Detection Sequence			What Gets Logged	What Might Actually Be True
Server looks for cookie file	Is cookie file present?	Yes, for User A	This is User A, a repeat visitor	This might be User B, a new visitor using User A's computer
		No	User is a new visitor	User might be User A, a repeat visitor who deleted his/her cookie file

Cookie files have received a bad name from instances where they stored information that invaded the user's privacy. For this reason, some users set their browser not to accept cookies, and log files cannot track these individual users unless their unique identity is available through the host address.

Path data

To follow a user's path through your site, you need to know:

~ Where the user entered the site

~ The sequence of pages the user followed

~ How the user moved from one page to the next

~ Data the user supplied as part of interacting with the site

~ Files the user downloaded from the site

~ Where the user left the site.

Because the log file records all user behavior as it occurs, lines representing one visitor's interaction with the site are interspersed among lines for all other visitors active during the same time. If each line provides enough data that you can distinguish one user from another, you can track a user's behavior.

Episodes vs. sessions

Many Web sites identify user "sessions" in their own exchange of data with the user. This session-ID data may be used in directory and page naming for dynamic page construction. Or it may be used in constructing "shopping carts" and similar cumulative interactions with the user.

Normally, these "sessions" are not completely adequate in usability analysis. For example, a site-defined session might begin when a user accesses the home page. However, a user might return to the home page many times as part of a looping-back strategy for site navigation (Catledge et al). Usability engineers want to know about use of this strategy because it may indicate a design deficiency.

Rather than rely on the site developer's definition of "session," it is best to reconstruct what Tec-Ed terms "episodes" that capture user behavior from opening page to exit page. An episode is generally based on all exchanges with a specific IP address (or other means of identity), from first transmission up to a gap of at least XX minutes, where XX is a number decided by the usability engineer. The gap length XX can also be adjusted depending on whether the log file indicates that the "next" page asked for after the gap was referred to from the page looked at right before the gap. The XX value should always be longer than any time thresholds being used in the software's own construction of "sessions."

Here is a simple illustration of the difference between session and episode construction:

Event Triggering a Log File Entry	Session Counting	Episode Counting
User's first access to the Web site	Session 1	Episode 1
User accesses other pages	Session 1	Episode 1
User returns to home page	Session 1 ends, Session 2 begins	Episode 1 continues
User accesses another page	Session 2	Episode 1

By constructing usability-oriented episodes, you can identify usability problems that may otherwise be masked by the rules used by the software to construct "sessions." For example, when home page access is used to identify the start of a session, the use of looping-back navigation to the home page cannot be analyzed using session-based data. Only by using episodes can you figure out what users are trying to accomplish with looping-back navigation, allowing you to devise other navigation solutions to help them achieve their goals more directly.

The episode construction described above is not completely appropriate for sites that expect extremely high-volume use from multiple users of a small set of IP addresses assigned on a DHCP basis or for IP masking in firewall security systems using proxy addresses. So major portal sites and sites expecting large numbers of simultaneous or near-simultaneous users from specific corporate user locations will need to use a more complex system for episode construction.

In some cases, explicit use of the referring-page and next-transmission links may be useful. In other cases, potential security problems (rather than usability issues) have caused site developers to encourage the use of an explicit logoff or sign-off procedure in (re)defining sessions. If this is done, the redefined session data may be useful in log file analyses.

Microsoft found this strategy appropriate in its features after receiving security criticisms based at least in part on the definition of a user session adopted implicitly in its software.

Naming for easier tracking

Determining which pages a user is visiting is simpler when page names are concise and self-descriptive. For many sites, dynamic pages constitute the bulk of user activity, yet their names (and the names of their components) make identification of their contents from log files almost impossible. You want identification of as much of the page content as possible to be visible in the log file.

For example, if all displays of catalog items are put into a standard page named response, or if each has its own unique name, it is difficult to tell if user failures to order some items are associated with the items being out of stock. To get at this type of information, you can name the page using a small, fast-transmission graphic with a detectable name to distinguish an outcome such as out-of-stock cases; the graphic should be named transparently—for example, ostock.gif, not pic002476.gif. Another possibility is to use daily or other-period stock reports from another source merged with log file information.

Other tracking-related data

To track the user's experience on the site for usability analysis, the log file must also contain:

- ~ A transmitted byte count (always logged) that can be used to detect searches that return no contents.

- ~ The contents of entries made by users in forms, both fill-in and multiple-choice-type entries. This information helps you analyze both what users are attempting to search for or request that you may not have made available, and identify usability problems. For example, misspelling may be an issue. You may find that users are entering "motors, electric" when the site reacts well only to "electric motors" or "motor, electric." In many cases, even the site's developers may not be aware of the site's detailed behavior until you point it out to them by first finding log file episodes you don't understand and then replicating them yourself to determine what must have happened.

~ Error logging and information about transmissions that are "stopped." This information should be in the main log file, not in a separate file.

~ Information about referring pages. This information should be in the main log file, not in a separate file.

~ Information about browser configurations. The amount of information a site's technical staff can provide will vary. In some cases, you can get not only what version of the browser is being used but also such details as whether the browser accepts your cookies (but not, of course, whether or how long they are retained), and what type of Java or ActiveX programming it accepts.

You want all you can get of this information. Although you normally identify usability problems from other data in the log file, you may occasionally look at browser-related information to understand where or how its specifics contribute to the problems.

Time data

Response time data—both site-response times and user-response times—provides clues to many aspects of usability. Log files provide a source of observations of such response times. (For information on the mechanics of using log file data to produce response times for specific sequences of page transfers, see *Web Site Stats* by Rick Stout. Stout shows how graphics transfers, often eliminated from log files before analyses, can be used in this process.)

Site- vs. user-response times

Slow site response can indicate an overly large file. If transfer of this file is commonly interrupted (recorded in the log file as an error), then you can ascertain that visitors are not patient enough to view the file. You can use this information to improve Web site usability.

In contrast, a Web page that has a high average "user response time" (viewing time) very likely has content of great interest—or confusion—to visitors. To explore which case is true, you can use the log file analysis to direct further usability research.

Analysis approaches

You can't do much to improve the way time data is recorded in the log file, but you can choose an analysis technique that makes better use of that data.

Mean times. Analyzing user-response time data by calculating mean, or average, times can create misleading results. (The mean is commonly derived by adding up a set of values and dividing by the number of values in the set.) Real observational response times include a small fraction of extremely long delays, the causes of which have no relation to the usability of the site. For example, some users go to coffee or lunch between one page reference and the next. This behavior inflates the mean times; more troublesome, the amount of inflation is highly variable simply due to random sampling effects. Together, the inflation and its extreme random variation—even in reasonably large total sets of observations—make mean response times less than reliable for measuring usability-related issues.

Median times. A more effective approach to analyzing user-response times is to use median times. (The median of a set of numbers is the value such that one half the numbers are less than it is and one half are greater.) The median is unaffected by the presence of a small fraction of large values (or small ones, as may occur when some sequences are produced by various forms of machine retrieval of sites or large sets of pages). Measurements of medians are much less subject to variability than those of means.

For example, let's imagine the following visit durations to a hypothetical transactional Web page for the following numbers of visitors:

Visit duration in seconds	0-10	11-20	21-30	31-40	41-50	51-60	61-70	...	...	3001-3010
Visitors	83	146	36	8	2	0	0	0	0	1

If we used the mean visit duration as a measure, we would say the average visit length is close to 25 seconds. However, if we use the median visit duration, the average visit length is 15 seconds. If we drop the unusual 50-minute duration, we see the effects of unusual items on the mean, which immediately drops by about ten seconds, while the median is unchanged.

Although statistical methods based on means are more common than those based on medians, methods for using medians exist for any mean-oriented analysis. These are described in many statistical texts and handbooks and supported by many statistics computer programs. Or you may choose to take a course or spend some time with a consultant statistician to develop familiarity with median-oriented methods.

Percentiles other than the median. In some circumstances, a percentile other than the median is preferable. For example, suppose you are interested in the behavior of regular users but you don't have user login data or cookie files for determining regular users. As an alternative, you can use the 10th percentile or some other percentile that is likely to represent the faster users, who are most likely to include those most familiar with the site.

For example, in the above example of visit durations to a transactional Web page, the duration for the 10th percentile of visitors (60 visitors) who spent the least amount of time on the page was 1 minute.

Choosing the exact percentile—10 percent, 25 percent, or even 5 percent—may initially require some experience with analogous sites. You can also use log file data to improve your estimates of an appropriate percentile by considering the fraction of users that are robots, the fraction that abandon interaction within three clicks, and so forth.

Using log file data to improve user success and satisfaction

Log file data suffers from two key shortcomings for usability assessments:

- Log files contain no information on the user's goal in visiting the site.

- Usability engineers cannot generalize from log file data as they can from data collected by performing controlled or randomized experimentation.

Many critics say that these and other weaknesses (some related to the unobserved use of stored or cached page content) render log files of no use in usability studies. Tec-Ed does not agree. These and other weaknesses simply prevent log file data from serving as the sole basis for any usability recommendation.

In fact, as the only easily accessible data about real use of Web sites, log files are extremely valuable. Usability engineers may use log file data to:

- Develop questions to be addressed with other techniques such as heuristic evaluation

- Develop hypotheses to be examined in other settings such as usability tests

- Test hypotheses that have arisen from other methods such as heuristic evaluation or usability tests.

In some cases, log file analysis may find multiple uses. A likely scenario for incorporating log file analysis within a usability evaluation program is as follows:

1. Design site

2. Develop prototype

3. Perform heuristic evaluation

4. Incorporate feedback into alpha site

5. Perform usability test of alpha site

6. Incorporate feedback into beta site

7. Perform log analysis of beta user activity

8. Use log analysis data to identify areas for further testing

9. Perform usability test of beta site

10. Incorporate feedback into first release

11. Perform ongoing log analysis of site

12. At intervals, use log analysis data to identify areas for further testing and improvement by developing questions.

For example, suppose analysis of log file data suggests that user response to a specific page is quite slow when compared with others. (After all, a log file cannot tell what the user is doing within a page, just from one page to another.) From this analysis, the usability engineer develops the question: "What causes the slow response, and how can we improve things for the user?"

The next step can be a heuristic evaluation of the specific page(s) involved. The heuristic evaluation might find that several elements of the graphic presentation seem to represent active links when they are not, and hypothesize that users are spending their time trying to click inactive elements. Simple changes developed to solve this problem can be tested by placing the modified page live on the site (or on one of several mirrors if the operation involved is large enough) and using log file data analysis to confirm that the original time problem has been reduced.

By testing hypotheses

As another example, suppose heuristic evaluation of the site's information architecture and navigation structure identifies areas where users seem likely to follow clumsy or overlong navigation paths that a new link structure might improve. The usability engineer can use log file data to determine if actual users seem to follow the clumsy or overlong paths, or if they navigate some other way to the target area. Or the log file data may show the target area to be underutilized, possibly because of the architecture problem identified by the heuristic evaluation.

To test some hypotheses, it may be necessary to look at historical data above the episode level, such as over a day, week, or month. This type of analysis typically requires user login or other identity data.

Conclusion

Assessing Web site usability from log file data is a new and evolving field. If you're performing this type of usability assessment for the first time, you'll need to work closely with technical staff who have expertise in the software and configuration choices used in producing the log file. You may also want to seek advice from, or engage in your project with, someone who has actually examined usability questions in a log file context. As in any dynamic new technical field, spending a few hours with people who have pioneered the technique can radically reduce the low-productivity weeks you'd spend toiling on your own—and accelerate the useful results of log file analysis.

References

Catledge, Lara D. and James E. Pitkow. "Characterizing Browsing Strategies in the World Wide Web." *Proceedings of the 1995 World Wide Web Conference,* Darmstadt, Germany, 10–13 April, 1995. (http://www.igd.fhg.de/www/www95/papers/80/userpatterns/ UserPatterns.Paper4.formatted.html).

Drott, M. Carl. "Using Web Server Logs to Improve Site Design." Association for Computing Machinery (ACM) *Proceedings on the Sixteenth Annual International Conference on Computer Documentation,* September 23–26, 1998, Quebec City, Canada, pp. 43–50.

Stout, Rick, (1977). *Web Site Stats: Tracking Hits and Analyzing Traffic.* Osborne/McGraw-Hill, Berkeley, CA.

Wilson, Tim. "Web Traffic Analysis Turns Management Data to Business Data." *TechWeb,* April 2, 1999. (http://www.internetwk.com/story/INW19990402S0006.)

Usability Evaluations for Commercial WWW Sites

Sue Braun,
Northwestern Mutual Life

If a goal of your site is to generate traffic and keep visitors, you must consider the WWW environment throughout your design and evaluation process. Potential users of your site have many choices of competitive sites to visit. There are few barriers to changing sites or incentives to stay on your site. Unlike productivity applications, users have not invested in long learning curves or proprietary data formats. The result is a competitive environment where users really are in control of whether and for how long they explore your site. In such an environment, usability is a key.

Text, graphics, interaction, and links

Whatever your user's purpose and type of site, your site is likely to be designed around significant text content:

- Many commercial WWW sites have evolved from print material. Content is often converted from existing print materials. Obviously, much about your organization, products, or services is described in print.

- Your site must provide (text) information for varied users. Your users do not have the benefit of a customer service or sales rep to analyze needs and provide only relevant information. Even if you collect demographic data from visitors to your site, it's probably not cost-effective to hone in on providing only a small amount of information to a narrow user profile.

Text (and graphics) present information chunking and labeling challenges not always found in productivity applications. On the other hand, WWW sites typically do not include the interaction complexities presented when users must manipulate complex data relationships. As a result,

• Site content must be oriented for the new or infrequent user, so it can't be overly complex

• The technology doesn't support complex structure.

However, the high-level interactions represented by intra- and inter-site linking strategies are extremely important on WWW sites.

The role of usability and HCI

Your multidisciplinary design team needs to know the target user's purpose and usage patterns and needs to design and evaluate in iterations. Therefore, HCI (human-computer interaction) and usability methods are as useful in WWW site design and development as they are in productivity application design and development. However, HCI and usability methods do not replace a content expert, information designer, graphic designer, editor, or other individuals skilled in communication-oriented design.

Contextual research

The combination of narrow functionality for a broad user profile, existing print content, and tight deadlines may mean that contextual user research is not the most effective use of the team's time. Hopefully some traditional market research has been performed. If not, the face-to-face nature of the usability lab makes it an indispensable tool.

Successful techniques: a case study

We have found the usability lab to be a very good tool for measuring:

• Initial impressions stated orally and observed in facial expressions and exclamations

• Problems for novice users

• How well the user understands specific content.

The lab has also been a great introduction for team members not familiar with HCI or usability methods. It also has been our team's only opportunity to meet potential users face-to-face. This has been a great team-building experience—there is nothing like a struggling user to bring focus to issues among content experts, designers, and technical experts.

The following recommendations result from our experience designing and evaluating the Northwestern Mutual Life WWW site. Lab evaluations have spanned ten months. As long as a lab evaluation is viewed as an important milestone, I believe others would realize benefits from lab evaluations.

You don't need a high-tech usability lab

At Northwestern Mutual Life, we do have a well-equipped usability lab. But we also run scenario-based "talk aloud" lab-style evaluations in conference rooms. We evaluate paper designs in conference rooms and in the lab, and we evaluate online designs in conference rooms and in the lab. Either setting can be appropriate for both "informal" and "formal" evaluations.

Northwesternmutual.com

The site supports personal insurance planning, small business planning, and sales force recruiting. The user cannot buy anything online, and the games are intended to provide some customized information, not really to challenge the user. The goal of the site is to get users to contact a Northwestern Mutual agency.

Our productivity application designers know their users quite well. This was not the case with WWW design. The design team started with material originally developed for print and online help. Audience research and usability evaluation had been completed on some of the material when originally developed. However, significant research of the target users in the new medium was not planned. Because the usability lab is a highly visible and successful tool for us, time for lab evaluations was planned.

Document goals and structure evaluations accordingly

The scope of the evaluations was wide. We evaluated against 14 general design goals, addressing such questions as, "Do users receive a clear and positive initial impression?" and about 20 much more specific design goals.

Evaluate paper designs first

As with productivity applications, we have had good success evaluating paper designs. In addition to card sorts and graphics evaluation, we ran informal scenario-based "talk aloud" evaluations of paper designs.

We used printouts of proposed graphics. Where content was not yet available, we used "greeked" paragraphs (nonsense foreign language phrases that some authoring tools produce as place holders). While evaluating paper designs, which lacked the intended content, it was not as important to have evaluators with a strong interest in our content. These evaluators (who were easier to recruit) nonetheless gave us tremendous data on organization, navigation, and our usage of WWW conventions. We designed a paper prototype and organized and conducted an entire evaluation in about two days using two staff people and eight evaluators. Since we could not focus on content, we kept each evaluation to about 45 minutes, which was time well spent.

Quickly move online

As designs moved from conceptual to interactive, from paper to electronic prototypes to alpha code, we designed usability evaluations to simulate a progressively more realistic production environment. Do not wait too long to factor in the realities of download times in your WWW evaluations. User-based results bring focus to any team. In our evaluations some specific evaluator comments ("I like it, but it's just too slow!") gave tangible tasks to those charged with implementing graphic compression techniques. Other comments ("This is great information, but I wouldn't scroll through and read most of it!") gave tangible tasks to those charged with chunking and labeling text. We considered comments only in context of all results, but specific observations really have an impact.

The final implementation plan included maximum download times in a particular environment, as well as a style guide. The lab evaluation results shaped these deliverables.

Recruit evaluators interested in your purpose

We have documented profiles of users we believe will find our content useful and interesting. That made recruiting appropriate evaluators more focused. Nevertheless, guaranteeing that each evaluator is interested in the purpose and content of the site has been a major challenge.

At a minimum, make sure your screening questionnaire removes individuals who would not be expected to find your content of interest. Our screening process itself gave us insights into target users' likely purpose and appropriate content.

Include experienced WWW users as evaluators

Our goals necessitated experienced WWW users as evaluators. You may have goals which include evaluating the usability of your site for the novice WWW user. But if you don't, at least 50 percent of your evaluators should use the WWW at least twice a month. You'll want to know how using your site compares to other experiences on the WWW.

Locating and compensating evaluators

In some target groups (e.g., college students), WWW users are not difficult to find. To find students interested in sales internship programs, we worked directly with campus placement offices. WWW users were a large percentage of the pool of potential evaluators we reached. Compensation was $40 for a two-hour time commitment.

For another evaluation, we needed evaluators who own a small business or professionally advise small business owners. Searching the "Business to Business Yellow Pages" or another traditional source of business owners, would have yielded a small percentage of WWW users (we believe around 10 percent). Therefore, we worked with a market research firm, who in turn purchased lists from online information service providers. Lists indicated the subscriber's profession and yielded about 150 names of WWW users in a six-zip-code area who appeared to meet our target evaluator profile. In fact, the market research firm was able to recruit with nearly 50 percent of calls yielding a qualified candidate. Compensation was $70 for a two-hour time commitment.

With many usability evaluators the non-financial incentives are important, i.e., the opportunity to preview and affect the site or product. To counterbalance this, we did include several evaluators who were "friends of friends" and evaluated as a favor. That guaranteed we had some evaluators who were less motivated by seeing early designs (less WWW techie). These individuals had no close affiliation with Northwestern Mutual Life. That also could have skewed the results of the evaluation.

Make the network connections and graphics realistic

We created a realistic (not too slow, not too fast) environment. That included a 14.4 kbps modem dialing up to an external server through an Internet service provider. If you are evaluating under a similar situation, make sure the connection is very reliable. Connections that can't be made or are lost during the evaluations, although a part of the user's environment, probably do not further the goals of your evaluation. The disruption could cause a loss of the usability data the evaluation is designed to capture. Don't forget to clear the browser memory and page cache between evaluators!

Use co-discovery

When planning our evaluations, we double-booked the evaluators to guarantee a minimum of evaluations completed if some evaluators did not show. We used co-discovery when both evaluators showed up. The results were excellent. Evaluators doing co-discovery:

- Described their expectations for content in more depth;

- Described their reactions to graphics and style more openly. They talked about their enjoyment, or lack thereof. This is possibly because these are matters of taste, and taste seems more appropriate articulated to a person next to you, as opposed to on a formal survey or during a formal debriefing. Reactions to style may seem out of place to a lone evaluator talking out loud;

- Enjoyed articulating their opinions to another human being in the same room, possibly because WWW usage is so anonymous.

A problem with co-discovery can be that one evaluator's opinions overwhelm another, or that the evaluators explore the site differently than they might have on their own. These are probably not big concerns with WWW site evaluators since no one's reaction is likely more important than anyone else's. An exception would be where an evaluator not interested in the content (not a target user) overwhelms one with a strong interest in the content.

Consider replacing scenarios with cards

We have found it very successful to replace ordered scenarios on 8" x 11" paper with 3" x 5" index cards. Traditional ordered scenario-driven evaluations did not work well for us:

- We wanted the evaluations to be more user driven (like the WWW) and not pre-ordered by us;

- We wanted the evaluation to be more fun (like the WWW) as opposed to that of an office job;

- We had clear user-task scenarios, but they were short and simple;

- For trying to observe boredom or fatigue, an always available card like "have some fun, play a game" worked much better than the same scenario in an ordered packet.

How cards work

1. On individual cards, list a piece of content (informative text and graphics, interactive activity, link to other sites, forms, e-mail, things to download, etc.). Include "free cards!" which brings out ideas for additional content—"no more cards are interesting," "I'm linking to a new site," "Find something fun," and "Take a break— eat a cookie."

2. To simulate an environment in which the user is under no pressure to hang around your site, have available competitive sites as bookmarks. It's important to come to a conclusion as to how long a target user would stay with the design being evaluated. You can also collect very limited and inexact reactions to competitors' sites.

3. Randomly order the cards and spread them out on the tabletop so the evaluator can move between working with cards and using the site.

4. Have available a sheet suggesting the evaluator follow this general protocol.

   - Choose a realistic or interesting card

   - Read the card aloud

   - Explain why you chose the card

   - Use the site to answer the question on the card

   - As you go, explain if the content is:
     - easy to find
     - easy to scan
     - easy to read
     - what you expected
     - easy to understand

   – interesting
   – useful
   – reliable
   – complete. If not, what is missing?

- What would you do as follow-up?

5. Make it clear to each evaluator that he or she can

   - quit at any time

   - express a need for new content

   - do some activity outside of the site, such as visit another site.

6. Include any questions to be answered immediately on the back of each card.

Do not expect the evaluator to follow the steps verbatim. Use the typical strategies for getting the evaluator to think out loud, not blame him or herself, etc.

Start the evaluation from other than the main page

We run target print advertisements that reference a URL below the main page. We also link from other sites directly to that lower page. Therefore, when evaluating that content, we started the evaluation on that lower page.

We were interested in whether evaluators would discover the rest of the site and find anything interesting. We also have many links from the originally sought-out content to other sections of the site. Our design goal is for users to recognize these other parts of the site exist and be able to navigate quickly between different but related parts of the site. The more users do this, the more they learn about us, our products and services.

Track these things during the evaluation

Initial reactions to the site. After the briefing, have the facilitator stay for a few minutes. After the evaluator has viewed the first screen (640x480 pixels), ask for an initial reaction as to the purpose of the site. Ask whether graphics are recognizable and attractive. This reaction should be accurate and positive. If it isn't, your user could be uninterested or frustrated—then quickly gone. Ask what the evaluator would do first (index, site map, what's new, about this site). Determine if the evaluator can find this desired first piece of content. Although each evaluator will not choose the same first task, you'll learn about user preferences on your site.

Structure, navigation and orientation, index problems.

Reactions to specific content.

Reactions to style and speed.

Cards chosen before diversions (breaks, links off the site, or claims of "no more interesting content"). In the production environment, is the user likely to return after the diversion?

The order cards are chosen. Consistency between evaluators gives you information about which content to list at the top level and how to progressively disclose other information.

Cards not chosen. Cards never chosen indicate low level of interest with these target profile evaluators.

How long before the evaluator finds no more cards of interest. If it's not long, determine whether it's the content or the evaluator that is the problem. If it's the evaluator, does he or she really meet the target profile? If not, do your target users really exist in large numbers? How do you recruit them as evaluators? How do you attract them to your WWW site?

Made-up cards. These are ideas for new content or labeling changes for existing content.

Why cards are not chosen. It may be that the evaluator could find the content, but in reality, wouldn't. Or it may be that the label, not the content, is what was not interesting. Or perhaps the content can be removed.

Is there enough non-text content to keep users engaged? Are users bored?

These are excellent topics for debriefing and a written survey after the evaluation. We have found written surveys effective for tracking the evolution of our site. For example, during one evaluation, the written responses to the "image" of our site were consistently "fast, clean, and professional." When we subsequently added larger, slower graphics, the written responses excluded "fast," and included the description "slow and professional" repeatedly.

Gather marketing data during debriefing

The usability evaluations were the first chance our team had to learn about WWW usage directly from individual target users.

Debriefing sessions provided an opportunity to ask market research questions: What content is missing? How often do you use the WWW? For what purposes? Which browsers? Which modem speed? Which search engines? Which online advertisements? Which print media? How do you learn about new sites? This is not quantitative data and it complemented only more traditional research. Nevertheless, the answers gave us useful insight into design, content, and advertising strategies.

We also asked specifically how our site compared to other sites the evaluator had visited. Again, this data really only gave us ideas on which sites to visit ourselves. Since sites change so quickly, we have not spent money on formal comparative evaluation.

Expect a blurring between users and authors

Many experienced WWW users have authored and published their own HTML documents. They may comment on a badly aligned table or clumsy navigation quite differently than an experienced user of a traditional application might. For example, experienced word processor users don't write word processor code.

Over 30 percent of our evaluators viewed source code and started to articulate detailed ideas for improving the documents. At other times, they viewed source code to research something they considered a clever implementation.

Watch for this mentality and try to curtail it during the evaluation. You are paying your evaluator to act like a user, not a site designer. But that is difficult if they are indeed both. Debriefing would be a more appropriate place for discussion of technical authoring issues.

Lab evaluations have shortcomings

It is difficult to measure the entertainment value of a WWW site in the usability lab. As detailed above, we measured desirability and engagement more than entertainment value. We also were not measuring suitability as a challenging game, so we did not consider these lab shortcomings to be problematic.

If your site is large, even desirable content may not be used by any evaluator. Evaluate separately or rely on your usability experts and editors.

Consider differences in browsers

You may not find it effective to run a full usability evaluation with different browsers or with a non-frames version of a frame-based site. If not, make sure this is a QA step completed by an editor, designer, or QA professional. Differences between browsers can be tremendous, especially with tables and graphics. Complete the evaluation with the browser used by the majority of your target users.

Other evaluation methods

For a team new to WWW design, we have not found a substitute for watching an evaluator, face-to-face, in a traditional evaluation setting. With a base of traditional evaluations to build a core context for your team, you can then collect a higher volume of data using other methods.

Evaluate graphics in and out of context with the site

Focus on evaluating graphics a lot, initially outside of the context of content evaluation. Have a talented graphic designer design several graphics for evaluation. Ask evaluators to fit graphics to descriptions. Have evaluators orally state and fill out a survey on how recognizable and attractive the graphics are. What is the style? Iterate. Remember to move online quickly as attractiveness and communication value cannot be isolated from speed on the WWW.

Online focus groups

We have recently added online focus groups to gather a higher volume of information on what users expect to find on our site. We used an online service to help us recruit and organize the groups. Groups include established business owners, new business owners, and women business owners. We plan to use the groups to support design and evaluation by soliciting comments on information hierarchy ("Where would you expect to find a topic?"), menu sort, graphic recognition, and attractiveness.

Online beta evaluations

Online focus groups can also evaluate a beta site. Ask them to fill out a survey and follow up with a phone conversation to at least a few evaluators.

Share your knowledge

Congratulations! Your site is desirable, useful, and usable to a potentially large number of WWW users. So, how do these users know that? Your site owners and marketers will certainly appreciate learning your recommendations for sites to link to and from, which paper and online advertising to target, and which content to add.

In targeted advertising, consider citing exact pages of interest rather than directing visitors through your main page. Push for a consistent style between the advertisement and the WWW site. Remember— marketing affects design and design affects marketing!